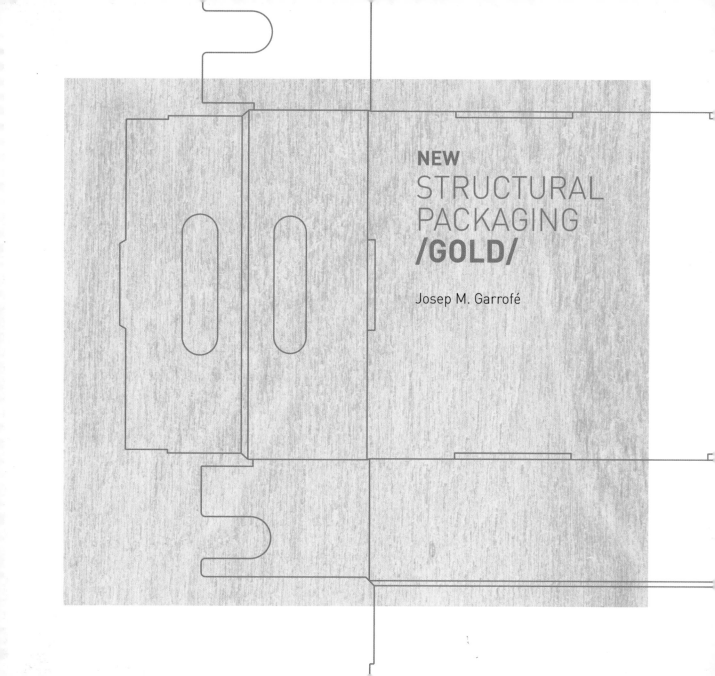

NEW
STRUCTURAL
PACKAGING
/GOLD/

Josep M. Garrofé

Hoaki Books, S.L.
C/ Ausiàs March, 128
08013 Barcelona, Spain
T. 0034 935 952 283
F. 0034 932 654 883
info@hoaki.com
www.hoaki.com

hoaki_books

NEW STRUCTURAL PACKAGING

ISBN: 978-84-17412-49-4
D.L.: B 1444-2020

© 2020 Promopress, Hoaki Books, S.L.
Second edition
© 2020 Josep Mª Garrofé
© all the photographs: Josep Mª Garrofé

Art Director: Josep Mª Garrofé / jmg@garrofe.com
Project Coordination: Laura Farré
Design assistants: Mireia Prats and Estel Roca

Printed in Turkey

http://structuralpackagingblog.com

We invite you to visit our blog. A site where we'll exchange structural packaging knowledge. In addition, you'll find a section to download different die-cut templates and monthly news.

FREE

Download
of all the projects

Visit our website www.promopresseditions.com.
Click on the tabs "downloads" and "extra free material", and enter the code **ZB9WQACCMY** for free access to all the diagrams contained in this book. By downloading this material you are registering into both Promopress, Hoaki Books database and also in the blog Structural Packaging database.
The diagrams' numbers correspond to those in the book and are classified according to the chapters. All the diagrams have been created by the author and are protected by copyright and intended for private use only.
The displays included in Structural Displays have been made with the materials, dimensions and techniques specified for each of them. Tribu-3 is not responsible for the results that may arise from changing any of its original conditions, especially if the appropriate tests are not made before the die-cut process.

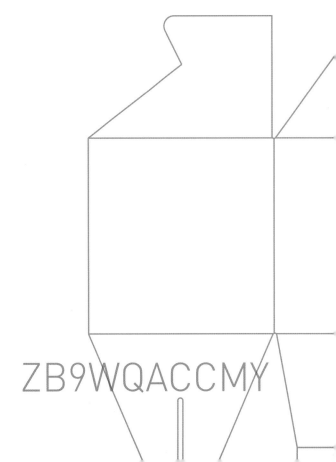

ZB9WQACCMY

INDEX
/SPG/

INTRODUCTION	06
TECHNICAL PICTOGRAMS	10
THEMATIC PICTOGRAMS	11
LEVEL 1	**12**
LEVEL 2	**110**
LEVEL 3	**250**
THEME INDEX	406

LEVEL1
/SPG/

012

001_Rigid accordion bag 002_Star-shaped surprise box 003_Picnic box/bag 004_wedding Gift box 005_Pillow box 006_Heart box 007_Triangle surprise 008_Double front package 009_Portable belt for flowers 010_Rabbit face pillow box 011_Double pyramid box 012_Simple rounded bag 013_Triangular case. 014_Eva foam bag 015_Chinese noodle box 016_Elephant candy box 017_Double sleeve cube 018_Document holder 019_Rigid tray with handles 020_Valentine's pack 021_Halloween bat with eyes 022_Candy cone 023_Four triangle box 024_Box for cakes 025_Arched base box 026_Simple cupcake box 027_Cup holder structure 028_Carrot box 029_Box for food 030_Flower basket 031_Desserts take away box 032_Christmas tree pack. 033_Cardboard heart for bottle 034_Ball inserted in a cross 035_Booster pack 036_Four-bottle structure 037_Batman box 038_Perfume diptych 039_Surprise box with "Ears" 040_Foam box for usb 041_Triangle with triangular closure 042_Gift box with cord 043_Car-shaped box 044_Matryoshka box 045_Two embedded triangles 046_Surprise pack 047_Tied-together triangles 048_Gift bag with handle.

LEVEL2
/SPG/

110

049_"Twisted" box for bottles 050_Welcome pack with chocolates 051_Software triptych 052_Double-layer corrugated cardboard sleeve 053_Cylinder for credit cards 054_Foam triptych box 055_Interlinking sleeves 056_Curved plastic packaging 057_Six triangle cube 058_Multi-layer sample box 59_Two interweaved triangles 060_Two-color surprise box 061_Block for cd 062_Two tray pack 063_Book with grass 064_Multi-use box-bag 065_Multi-layer cd triptych 066_Triple display box 067_Note pad folder 068_Grass between blocks 069_Three layer pack 070_Gift box for bottles 071_Box with raised base 072_Gift case with sleeve 073_Advent Calendar box 074_Ingot-shaped box 075_Product on display 076_Sommelier display pack 077_Translucent connected pack package 078_Triple bucket pack 79_Triptych display box 080_Ecological christmas pack 081_Two interlocking "l"s 082_Foam box with handle 083_Minimalist display case 084_Double symmetry box 085_Double cube pack 086_Gift kit sample case 087_Perforated square pack 088_Framed gift box.

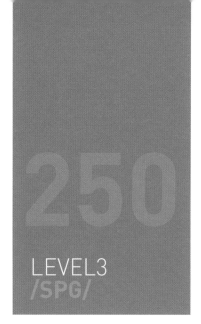

LEVEL3
/SPG/

250

089_Car welcome pack 090_Rounded-edge chest 091_Mobile display frame 092_"L"-border box 93_Multiple accordion box 94_ Diagonal surprise pack 95_Feminine fragrance welcome pack 96_New fragrance display 97_ Display tray welcome pack 98_Christmas tree box 99_Coffee maker display pack 100_Methacrylate box 101_Promotional case pack 102_Display tray insert 103_Plush welcome pack 104_Women's perfume chest 105_Encased bottle pack 106_Globe cube 107_Pack with curved corners 108_Sloped box 109_Credit card pack 110_Wooden container box 111_Personalized facial treatment 112_Rigid triangle in motion 113_Double fragrance welcome pack 114_Transformable cube box 115_Bewitching display box 116_Popcorn surprise box 117_Passe-partout book pack 118_New perfume pack 119_Book frame 120_Sloped welcome pack 121_Padded display case for glasses 122_Triple lid box 123_Sommelier gift pack 124_Wooden display frame 125_Luminous welcome pack.

/SPG/

...CREATIVE
RESTLESSNESS, THE
DESIRE TO DISCOVER NEW
PATHS, EXPLORE NEW
POSSIBILITIES AND TAKE
RISKS, IS PART OF OUR
TEAM'S DNA

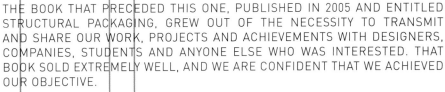

THE BOOK THAT PRECEDED THIS ONE, PUBLISHED IN 2005 AND ENTITLED STRUCTURAL PACKAGING, GREW OUT OF THE NECESSITY TO TRANSMIT AND SHARE OUR WORK, PROJECTS AND ACHIEVEMENTS WITH DESIGNERS, COMPANIES, STUDENTS AND ANYONE ELSE WHO WAS INTERESTED. THAT BOOK SOLD EXTREMELY WELL, AND WE ARE CONFIDENT THAT WE ACHIEVED OUR OBJECTIVE.

IN THE TEN YEARS THAT HAVE PASSED SINCE THEN, CHALLENGES HAVE NOT STOPPED KNOCKING AT OUR DOOR. IT IS VERY EASY TO REAP THE FRUITS OF ONE'S PREVIOUS EFFORTS AND REPEAT DESIGNS THAT WORK WITH A FEW SMALL VARIATIONS. HOWEVER, CREATIVE RESTLESSNESS, THE DESIRE TO DISCOVER NEW PATHS, EXPLORE NEW POSSIBILITIES AND TAKE RISKS, IS PART OF OUR TEAM'S DNA. IT ALSO SEEMS TO ME THAT MORE THAN 25 YEARS OF EXPERIENCE ALLOW ONE TO ASSUME THAT A CERTAIN DEGREE OF CREATIVE MATURITY HAS BEEN REACHED, IN MY OWN CASE AS WELL AS IN THE CASE OF ANY PROFESSIONAL THAT MUST CONTINUE CREATING AND LEARNING. IT IS FOR THIS REASON THAT WE WANTED TO PUBLISH A NEW BOOK THAT COLLECTS OUR BEST NEW DESIGNS OF THE LAST TEN YEARS, ONE WE ARE PROUD TO CALL NEW STRUCTURAL PACKAGING GOLD. IN IT, THE READER WILL FIND 125 PROJECTS, EACH THE PRODUCT OF OUR ONGOING LEARNING AND EVOLUTION.

IN A SECTOR AS STABLE AS PACKAGING, ONE IN WHICH THE RAW MATERIALS AND PROCEDURES BARELY CHANGE, IT IS DIFFICULT TO FIND INNOVATIVE STRUCTURES THAT CONTRIBUTE NEW IDEAS, ESPECIALLY WHEN THE RACE AGAINST TIME IS ALSO A FACTOR. AS A RESULT, THE NEED TO GO A STEP FURTHER AND COME UP WITH NEW DESIGNS FOR EACH PROJECT IS THAT MUCH MORE ESSENTIAL. AND THIS IS WHAT WE HAVE COMPILED HERE: FRESH PROPOSALS AND ORIGINAL IDEAS FOR EACH NEW PRODUCT, NEED OR EVENT.

THE BOOK CONSISTS OF THREE SECTIONS THAT GROUP PROJECTS TOGETHER ACCORDING TO THEIR DEGREE OF COMPLEXITY, FROM THE SIMPLEST PROPOSALS TO THE MOST SOPHISTICATED, WITHIN A WIDE RANGE OF SECTORS AND APPLICATIONS. EACH PROJECT IS UNIQUE AND IS PRESENTED WITH ALL THE DETAILS AND INFORMATION THAT READERS NEED TO BE ABLE TO REPRODUCE THEM ON THEIR OWN. HOWEVER, OUR ULTIMATE HOPE IS THAT THE MATERIAL SERVES AS A SOURCE OF INSPIRATION, OPENING FOR READERS THE MAGIC WINDOW OF CREATIVITY THAT ALLOWS THEM TO ADAPT THE DESIGNS TO THEIR SPECIFIC NEEDS, TO MODIFY THEM, PLAY WITH THEM AND TAKE ON THE CHALLENGE THAT MOTIVATES OUR TEAM EVERY DAY: THAT OF DISCOVERING NEW FORMS AND STRUCTURES. WE ARE DRIVEN BY THE DESIRE TO GO BEYOND ESTABLISHED BOUNDARIES, TO ACHIEVE NEW OBJECTIVES, AND WE HOPE THE BOOK TRANSMITS THIS SAME CREATIVE CURIOSITY TO ITS READERS AND USERS.

WE ALSO WANTED THIS BOOK TO BE A TRUE COMPILATION OF OUR BEST PROJECTS. IN ADDITION TO NEW PROPOSALS, INSIDE YOU WILL FIND A LINK TO AN ADDITIONAL SELECTION OF PROJECTS INCLUDED IN PREVIOUS PUBLICATIONS OF OUR WORK. YOU WILL SEE THAT EACH ONE OF OUR IDEAS AND DESIGNS IS UNIQUE. WE ARE DELIGHTED TO SHARE ALL OF THIS WITH YOU. WE HOPE THAT WILL YOU ENJOY DISCOVERING AND CREATING AS MUCH AS WE DO.

TECHNICAL PICTOGRAMS

/SPG/

PROCEDURES USED AND USEFUL
CONSIDERATIONS

 Fiber Direction

 Interior Weight

 Precise Manipulation

 Electrical Circuit

 Magnet

 Gluing

 No Gluing

 Foldable

 Offset

 Silk-Screen Printing

 Cardboard Gluing

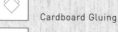 Lined

 Interior Reinforcements

 Half-Cut

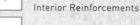

THEMATIC PICTOGRAMS
/SPG/

SECTORS AND ACTIVITIES IN WHICH
THE DISPLAYS ARE USED

 Pharmacy

 Technology

 Drinks

 Editorial

 Welcome Pack

 Children

 Accessories

 Nutrition

 Gifts

 Christmas

 Jewellery

 Cosmetics

 Merchandising

 Textile

 Sweets

 Perfumery

/18/ /74/ /20/ /14/ /22/ /24/ /70/

/32/ /34/ /54/ /100/ /40/ /42/ /44/

/50/ /52/ /90/ /58/ /30/ /62/ /80/

/60/ /84/ /72/ /26/ /76/ /104/ /98/

/92/ /16/ /86/ /82/ /36/ /88/ /94/

/56/ /96/ /106/ /78/ /108/ /102/ /38/

/28/

/68/

/66/

/48/

/64/

/46/

/LEVEL01/

RIGID
ACCORDION BAG

Use: Ideal for books or flat objects. **Development:** This bag is made of rigid die-cut cardboard with half-cut creases. The sides are made with zigzag-creased paperboard, forming an accordion pleat that facilitates opening the bag. The rigid cardboard can be silk-screened or laminated with paperboard. Keep in mind the creases and edges, as they will reveal the grey color of the cardboard. **Material:** 2 mm grey cardboard, 280 g paperboard. → 1618

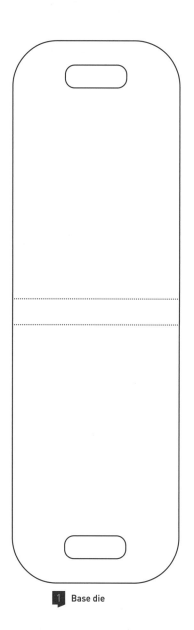

1 Base die

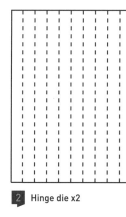

2 Hinge die x2

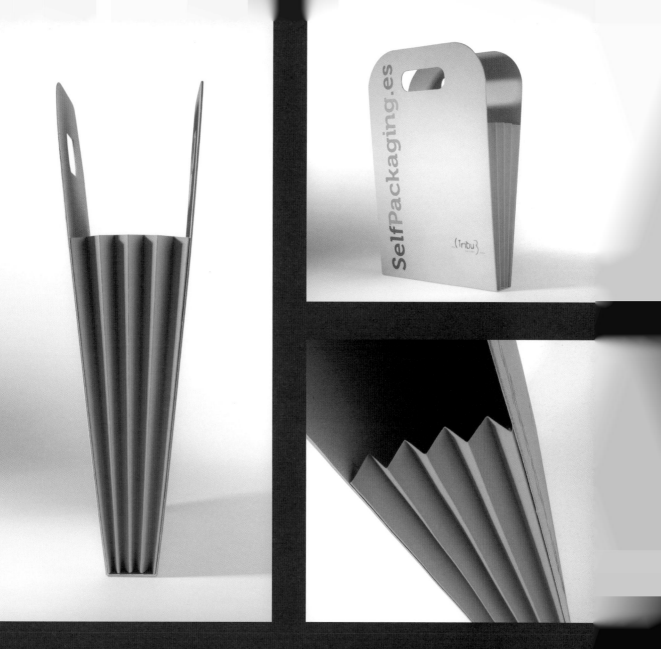

STAR-SHAPED SURPRISE BOX

Use: Decorative star for small gifts. **Development:** The structure is created with several pieces of double-layer corrugated cardboard glued together for thickness. The lid and the bottom are made with the same die in the shape of a star but without the inside circle so as to be able to close the box. We glue the pieces together and to the bottom, gluing a circle recovered from the excess material to the lid as well. This permits opening and closing the box with a lid that fits perfectly. **Material:** Double-layer corrugated cardboard, 18 mm satin ribbon. → 2234

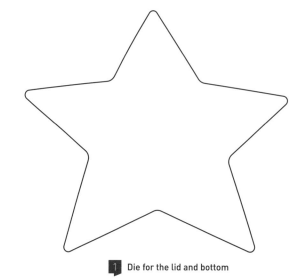

1 Die for the lid and bottom

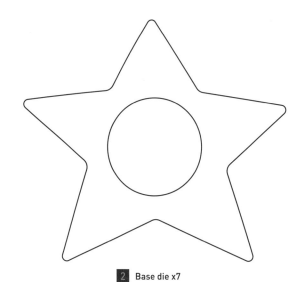

2 Base die x7

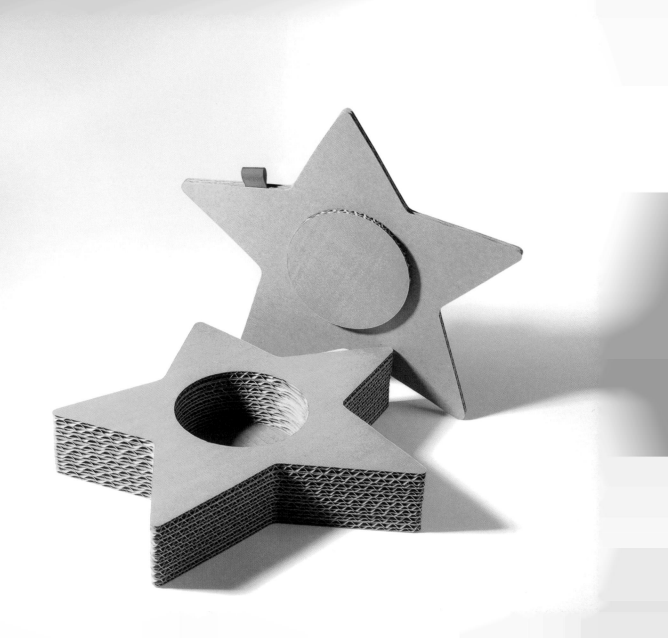

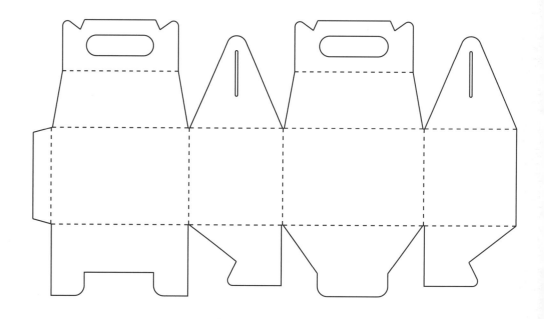

WEDDING GIFT BOX

Use: The perfect box for wedding gifts and small confectionery portions. **Development:** Oval box that closes by folding over its four strips. Joining these four parts creates a clasp that can also be used to easily carry the box via the die-cut circles. **Materials:** 300 g cardboard → 2272

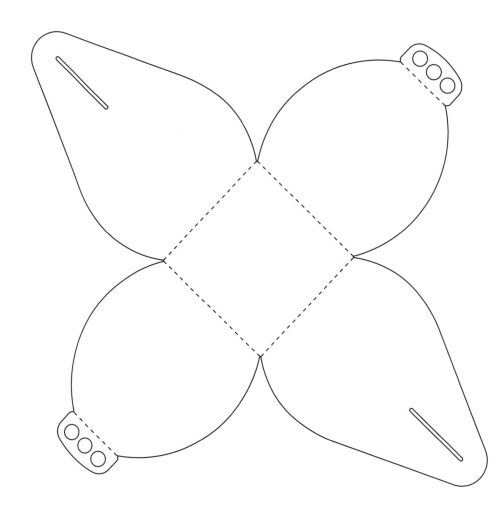

005

PILLOW
BOX

Use: A simple box with concave sides. **Development:** This box is made with a single die, giving its four sides concave curves. These permit easy closing of the box by applying pressure. This unconventional method of closing the box needs to be kept in mind when designing the die because the pack tends to open. It needs to be perfectly adjusted to ensure a tight closure. **Material:** 300 g cardboard. → 1614

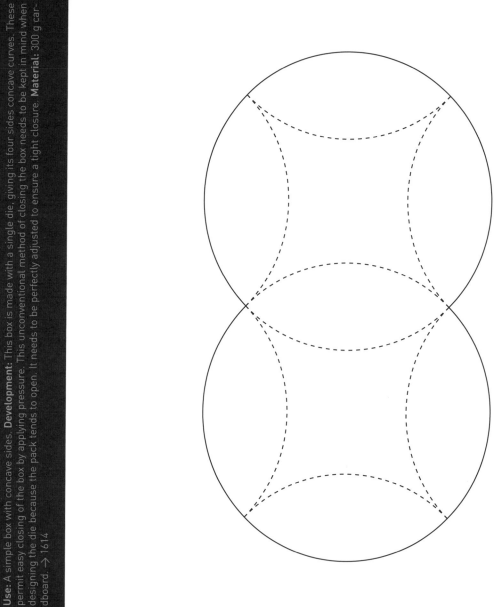

006

HEART
BOX

Use: Box with a heart-shaped fastener, perfect for Valentine's Day or wedding invitations. **Development:** This box has a simple fastener that forms a heart shape when closed and combines organic forms with the box's straight lines. **Material:** 300gr cardboard → 2524

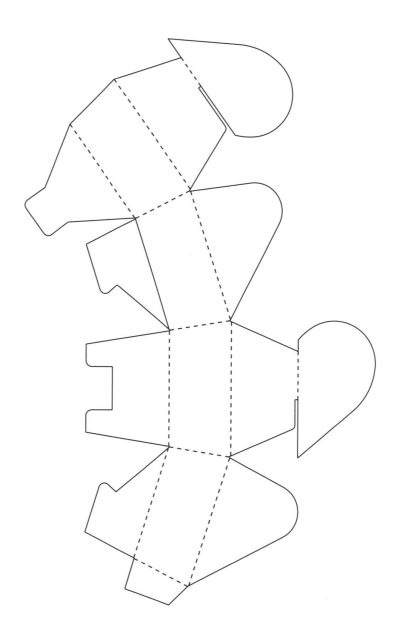

007

TRIANGLE SURPRISE

Use:: Box for small objects. **Development:** This box consists of a single piece folded into the shape of a triangle. If used for food products, the inside varnish needs to be suitable for these applications. The box is closed with an elastic band. A label can be added with the product name or information. Products in bulk will make the most of the space inside the box. **Material:** 280 g paperboard, 3mm Ø elastic band. → 0536

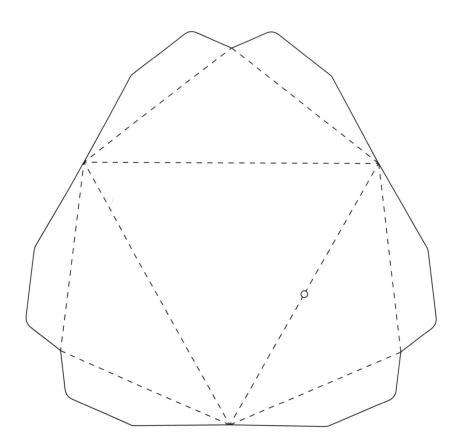

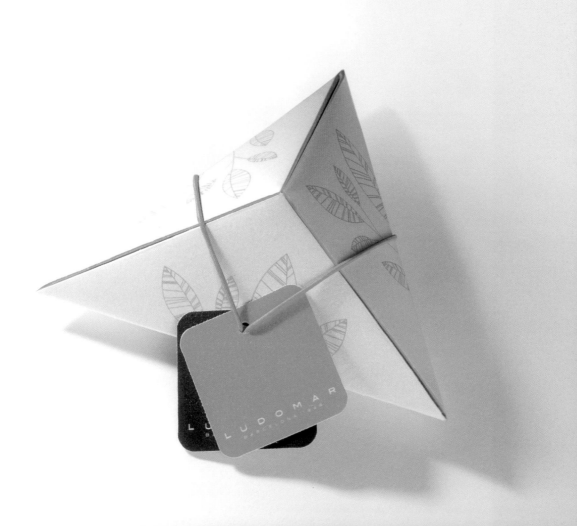

008

DOUBLE FRONT PACKAGE

Use: A box designed for nutricosmetic products that includes a double front that shifts towards the inside. **Development:** In order to achieve this double front, a simple single-piece die is designed that already includes the inside front. Its simplicity allows for industrial and mechanic production in large quantities. When assembling the box, the inside face shifts inside, thus providing an interesting double-front effect. **Material:** 250 g paperboard. → 2320

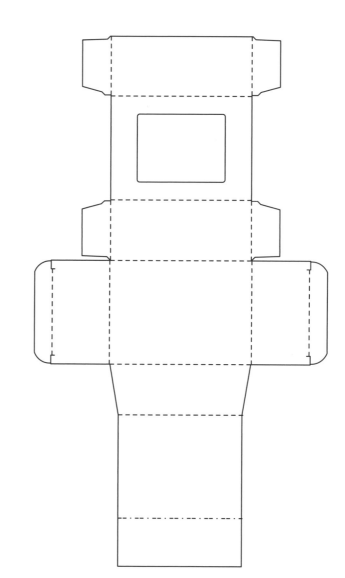

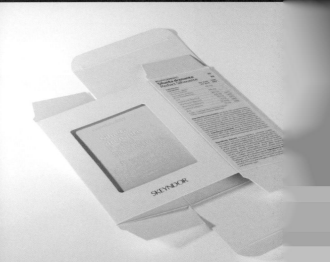

009

PORTABLE BELT FOR FLOWERS

Use: Box for transporting either an individual flower or small bouquet. **Development:** Box comprising three tabs that secure and protect the flowers in three places. Features die-cut circles that allow the box to be held by the fingers. **Materials:** 300 g cardboard → 2270

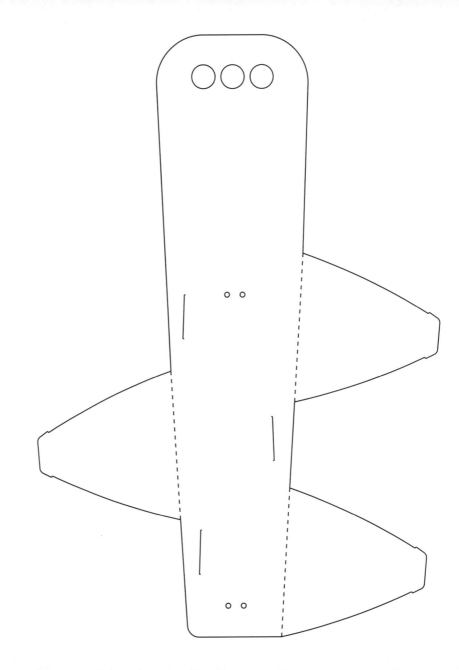

010

RABBIT FACE PILLOW BOX

Use: Simple pillow box with different designs. **Development:** Using a simple structure as the base, in this case, a standard pillow-box die, we add die-cut paperboard rabbit ears for decoration. The rest of the design is printed directly on the base die. This die creates an interesting curve, highlighting the illustration of the animal. Many other animals can be used for this packaging. **Material:** 300 g paperboard. → 2212

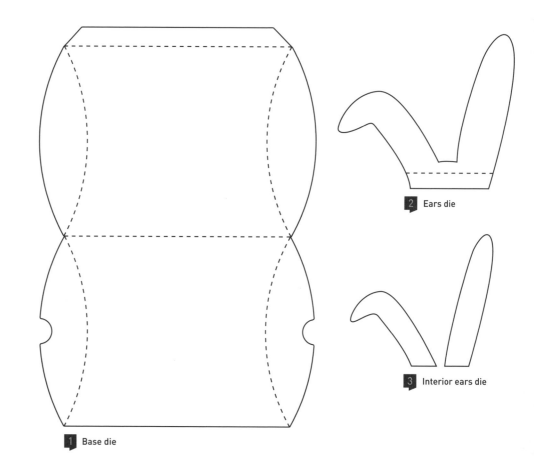

1 Base die

2 Ears die

3 Interior ears die

011

DOUBLE
PYRAMID BOX

Use: A unique presentation with an optical effect. **Development:** This box consists of two inverted pyramids, connected by the creases of the structure. This design results in a packaging with a double inside space created with a single die. When the top part is open, the two spaces become one. **Material:** 280 g paperboard. Box designed by ELISAVA [Alexandra Buxens] → 2171

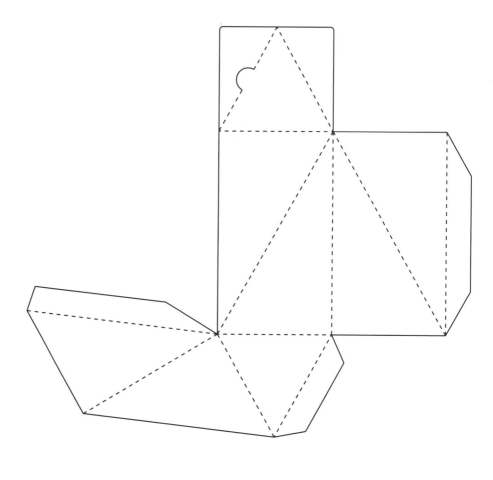

012

SIMPLE
ROUNDED BAG

Use: Simple pack for small gifts. **Development:** This bag is made with a single die. On the inside it's triangular, while on the outside the shapes can vary. In this case, we opted for a semi-circle. It is closed with two flaps included in the die. **Material:** 300 g paperboard. → 2202

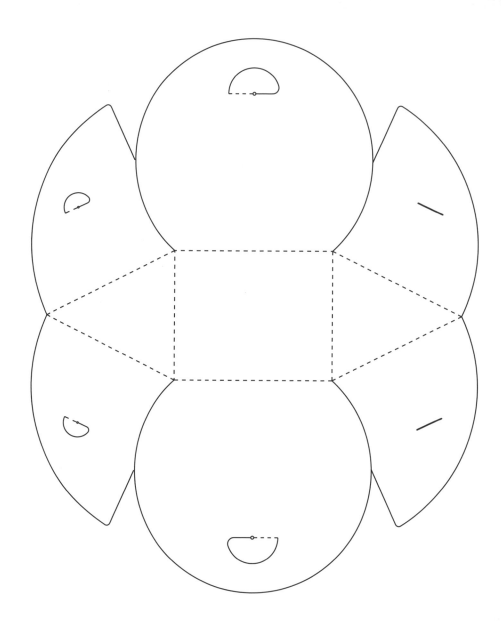

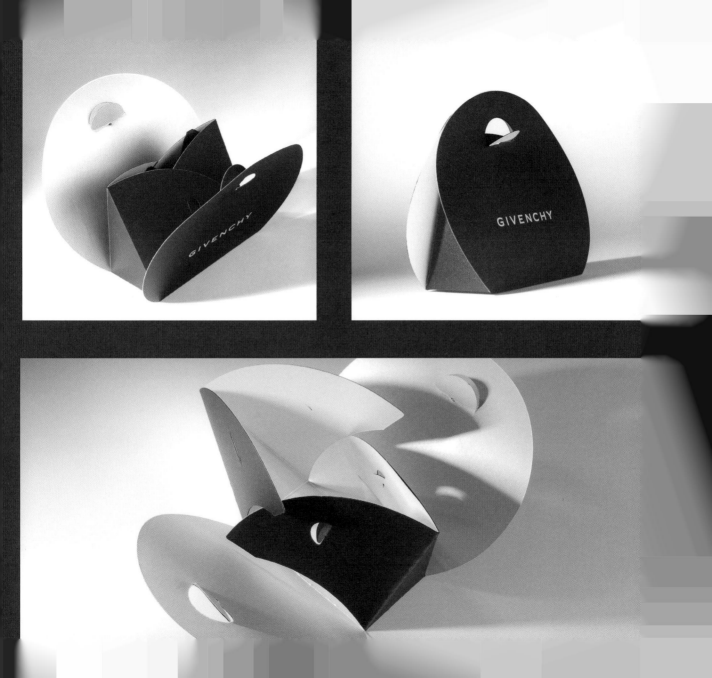

013

TRIANGULAR CASE

Use: Triangular case for small objects. **Development:** This apparently complex box is made from a single die that is folded to form this creative pack. The case acquires a triangular shape in which the flaps don't need to be glued to close the pack. **Material:** 280 g paperboard. → 1613

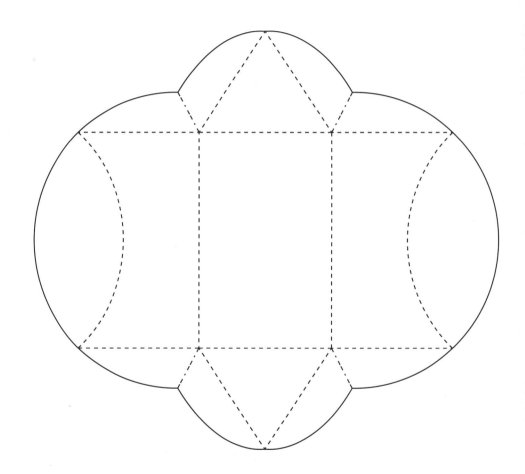

014

EVA
FOAM BAG

Use: EVA foam is for brochures and catalogs. **Development:** This pack consists of a single piece of EVA foam, folded to form a bag and which includes a hole for easy transport. The sides of the bag close securely with two spiral binders placed on the pierced material. It is ideal for flat, thin objects such as books or brochures. **Material:** 0.4 mm EVA foam, spiral binders. → 1057

015

CHINESE NOODLE BOX

Use: Originally designed to contain food products. **Development:** This simple box with self-assembly bottom is ideal for ready-to-eat hot food. In this case, using a heat-resistant material suitable for food products should be kept in mind. The handle, which can be made of string or metal, increases the uses of this packaging that, along with the self-assembly folding, facilitates storage in a small store of any kind, for instance, a clothing store. **Material:** 300 g paperboard, 6 mm Ø rope. → 2218

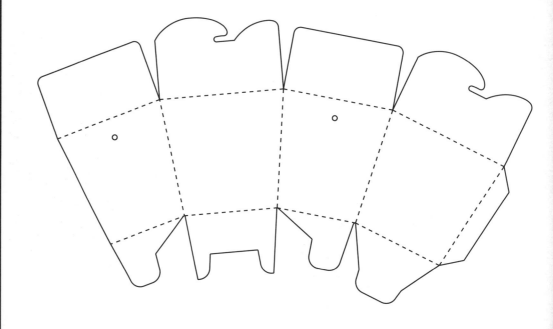

016

ELEPHANT
CANDY BOX

Use: Small box in the shape of an animal's head for kids' parties. **Development:** This original box is based on a very simple self-assembly structure. When you rotate the box, the lid becomes the main element, transforming into the cute head of an elephant. The lid can have different designs, thus offering multiple possibilities. A flap that fits perfectly between the animal's ears is used to close the lid. **Material:** 300 g paperboard. → 2217

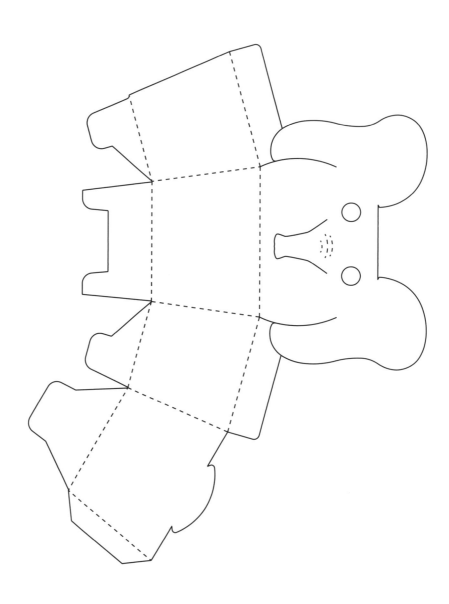

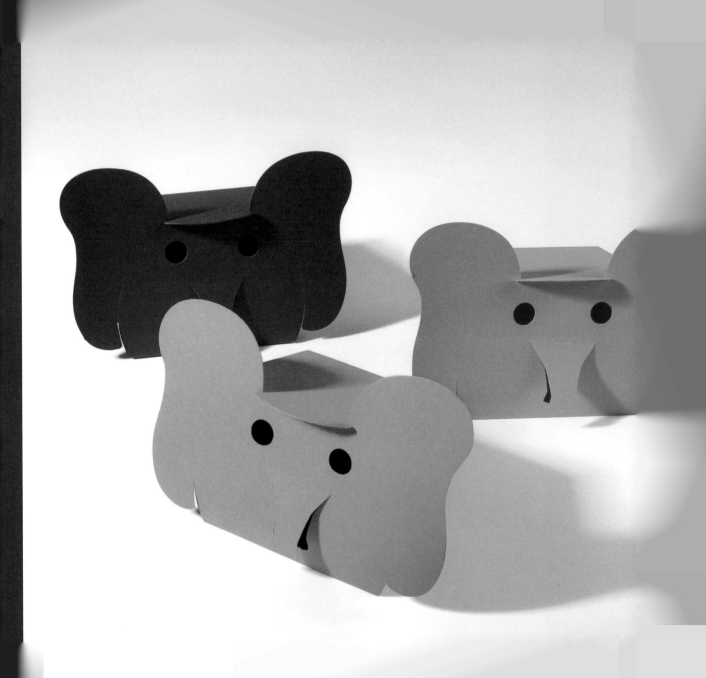

DOUBLE
SLEEVE CUBE

Use: Simple, folding box for small textile pieces. **Development:** For this box we used two sleeves to contain the buff, one made of cardboard and the other polypropylene. The exterior sleeve needs to be slightly larger than the interior one in order to fit properly. For this type of box, we can combine different materials. **Material:** 240 g cardboard, 0.4 mm polypropylene. → 0021

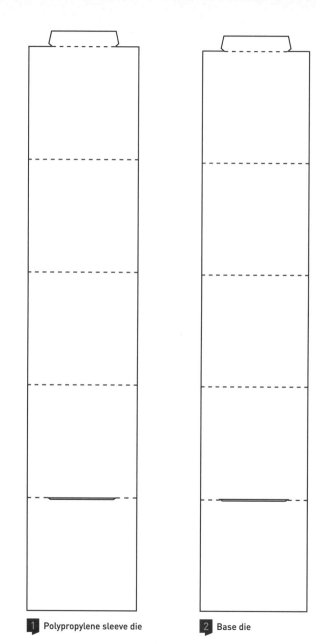

1 Polypropylene sleeve die

2 Base die

Use: Simple folder for events and presentations. **Development:** Only two simple dies are needed for this document holder. A piece with creases is glued to the base die (this is where the documents go). The folder closes with an elastic band. **Material:** 2 mm grey cardboard, 280 g paperboard and 15 mm Ø elastic band. → 1854

1 Lid and base die

2 Folder spine die

PLANIFICA LA TEVA AGENDA INTERNACIONAL AMB LA CAMBRA

PLA D'ACCIÓ INTERNACIONAL 10

019

RIGID TRAY
WITH HANDLES

Use: A tray for food products or bottles. **Development:** The structure consists of a single piece of micro-flute cardboard that, once folded, transforms into a strong tray. Because of the thickness of the material, a double crease is necessary. Another interesting detail is this simple die doesn't require glue when assembling the box. Simple flaps located on the base are all that is needed to assemble this piece. **Material:** 280 g micro-flute cardboard → 2211

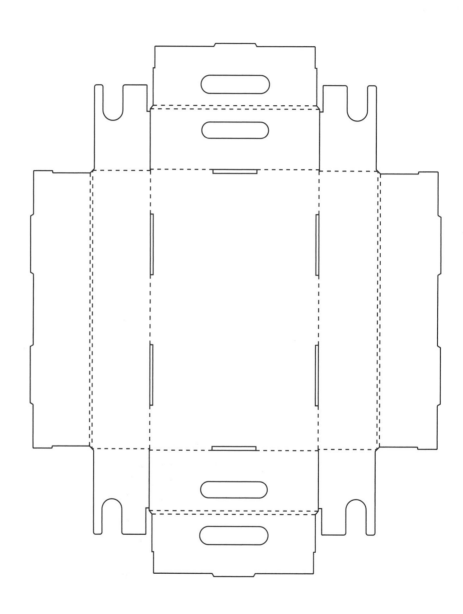

020

VALENTINE'S PACK

Use: Attractive heart-shaped diptych for product samples. **Development:** This simple but original packaging consists of a single die. One of the ends affixes to a slot in the other end, thus closing the pack and creating an aesthetically sinuous curve that covers the product it contains (chocolates, in this case), secured with flaps. **Material:** 300 g paperboard. → 1469

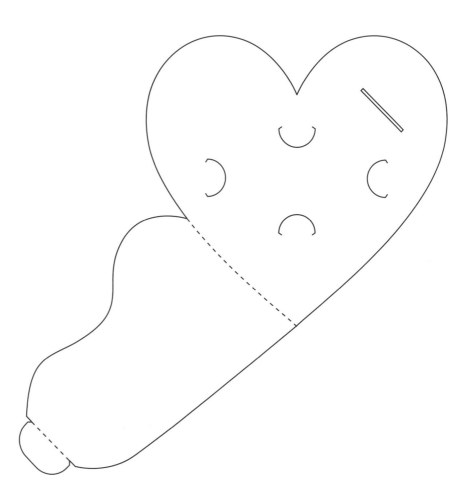

021

HALLOWEEN BAT WITH EYES

Use: Container for candy, popcorn, etc. Ideal for children's parties. **Development:** Simple self-assembly design, specially made for Halloween. Starting with a standard box without a lid, we add playful wings die-cut with the same paperboard material to the sides. We complete the design with adhesive eyes, and there you have it: a bat ready to be filled with candy. **Material:** 300 g paperboard, plastic adhesive eyes. → 2221

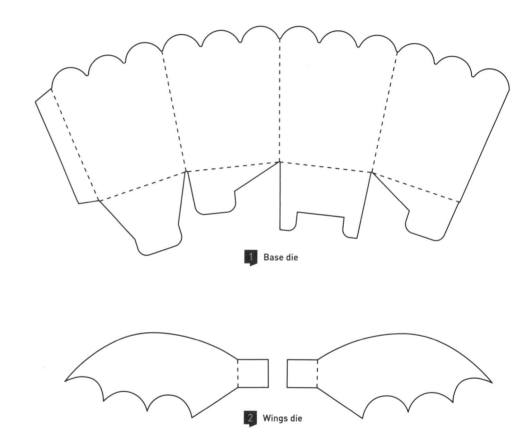

1 Base die

2 Wings die

022

CANDY CONE

Use: Long triangle box for informal gifts. Self-assembly. **Development:** This simple packaging was designed to be assembled quickly and without glue. Two flaps adjust perfectly into the slots on the opposite side of the die. Once the pack is assembled, these slots do not open. To close the pack, we interlace the two lids on the base. **Material:** 300 g paperboard. → 2267

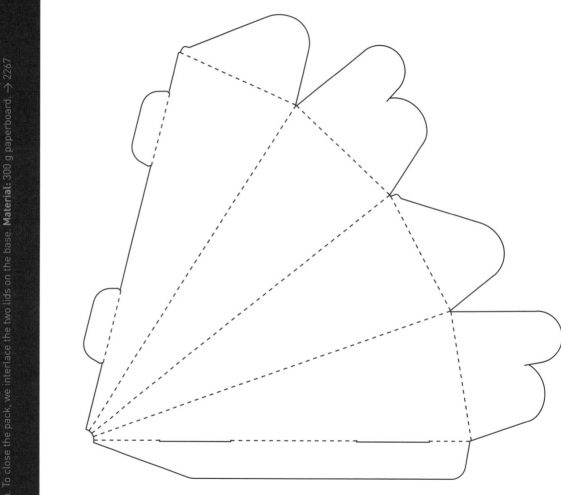

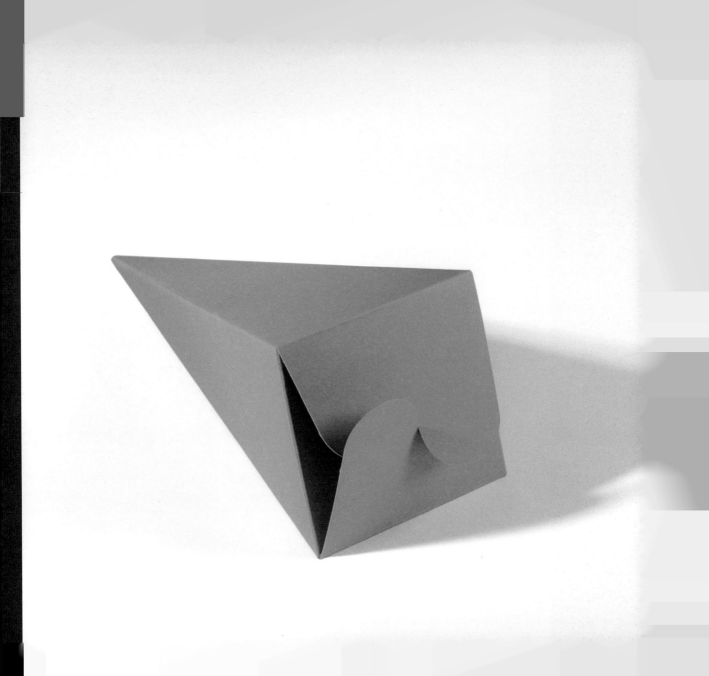

023

FOUR
TRIANGLE BOX

Use: An original way to present a gift. **Development:** The structure consists of four pyramids connected at the square base that interweave to close the pack. When you pull from the triangles, the box opens, revealing the inside. **Material:** 300 g paperboard. Box designed by ELISAVA (Gemma Sánchez). →2173

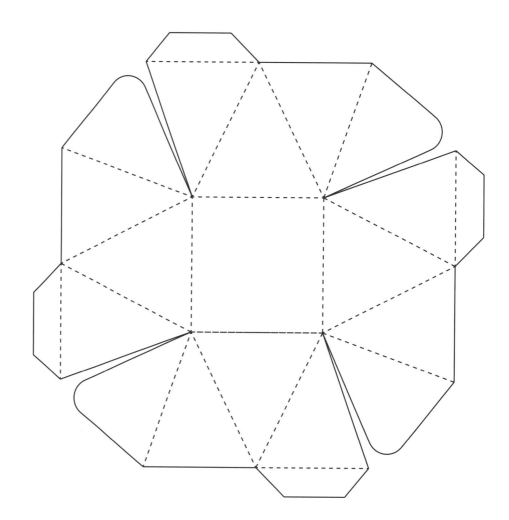

024

BOX
FOR CAKES

Use: Designed for easy transport of cakes. **Development:** This box was designed keeping in mind that it needs to be accessible from all four sides. We incorporate reinforced handles on the larger sides. The smaller sides can be enlarged with walls that easily cross over each other to form a simple cube. A transparent lid that reveals and protects the cake can be added to the pack. **Material:** 300 g paperboard. 0.3mm PET plastic. → 2242

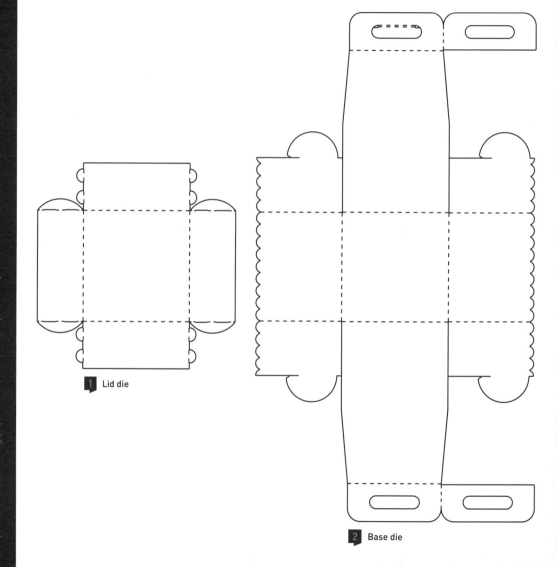

1 Lid die

2 Base die

025

ARCHED
BASE BOX

Use: A gift box with an arched base. **Development:** What makes this packaging unique is the oval arch on its base, creating a suggestive shape that distinguishes it from a conventional box. A regular die is used; only the oval arch on the base is changed. **Material:** 300 g paperboard. → 2332

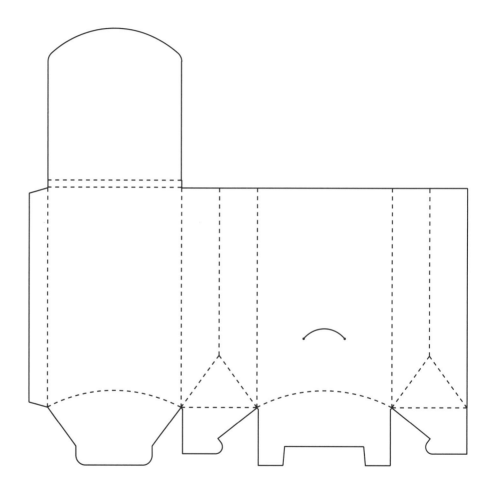

Júlia
PERFUMERIA

Felices Fiestas Merry Christmas Joyeuses fêtes
Bones Festes Boas Festas Buone Feste Frohe Feiertage
счастливых праздников 节日快乐

026

SIMPLE CUPCAKE BOX

Use: Specific for stores. It can be stored folded. **Development:** Easy assembly two-piece box that doesn't require any glue. The curves evoke the shape of a cake and the exterior lid, which doesn't require glue or adhesives either, is constructed with pressure flaps. An interior piece guarantees that the product is perfectly secured. **Material:** 280 g paperboard, 0.3mm PET plastic. → 3000

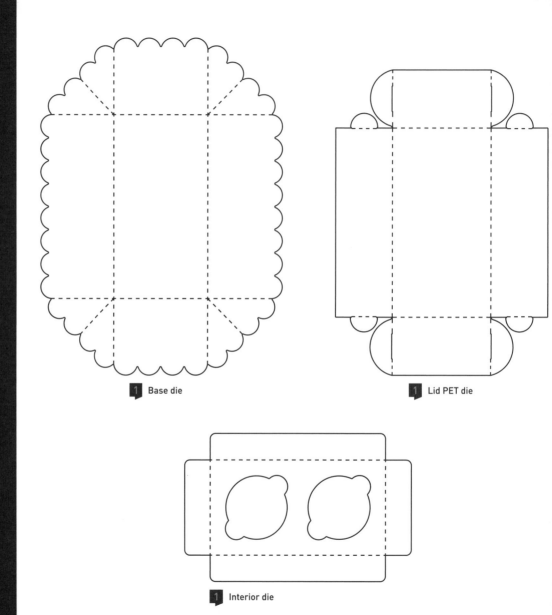

1 Base die

1 Lid PET die

1 Interior die

027

CUP HOLDER STRUCTURE

Use: Simple packaging to carry paper cups and avoid annoying burns. **Development:** Simple three-sided structure, with handles and a simple closing mechanism. We add holes to the base to hold the paper cups, as these are wider at the top. It is important to use thick paperboard as the base tends to bend due to the weight of the cups. **Material:** 300 g paperboard. → 2358

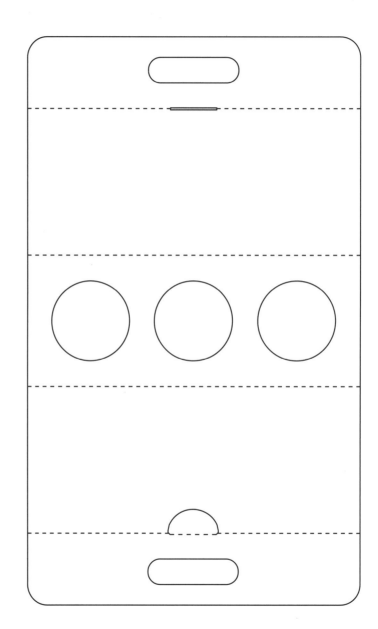

028

Use: Carrot-shaped box for Easter gifts. **Development:** Triangular box that is perfect for holding sweets and chocolates. Has a closable lid and features carrot colours. **Material:** 300gr cardboard. → CL23

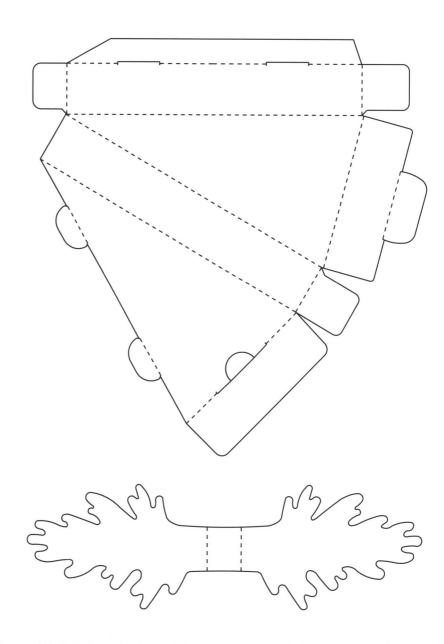

BOX
FOR FOOD

Use: A box with a large opening on top to make for easy removal of the content. **Development:** Simple folding structure with a fun handle for innumerable uses. The pack can be stored folded in small spaces and can be assembled instantly. It's ideal for stores that require speed. If used for food products, selecting appropriate materials or an approved laminate is required. **Material:** 320 g paperboard. → 2249

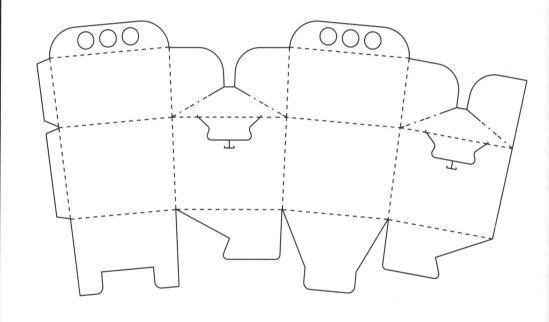

030

FLOWER BASKET

Use: Box with an opening for holding flowers and even displaying them vertically. **Development:** Semi-cone-shaped box with openings on both sides for holding and protecting bunches of flowers in a convenient and original way. **Material:** 300gr cardboard → 2513

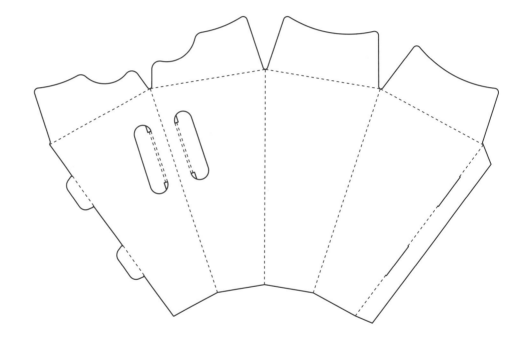

031

DESSERTS TAKE AWAY BOX

Use:: A box for cake slices that has been specially designed for businesses that offer take-away options. **Development:** A simple and convenient box that allows desserts to be transported and consumed in a take-away format. **Material:** 300gr cardboard. → 2517

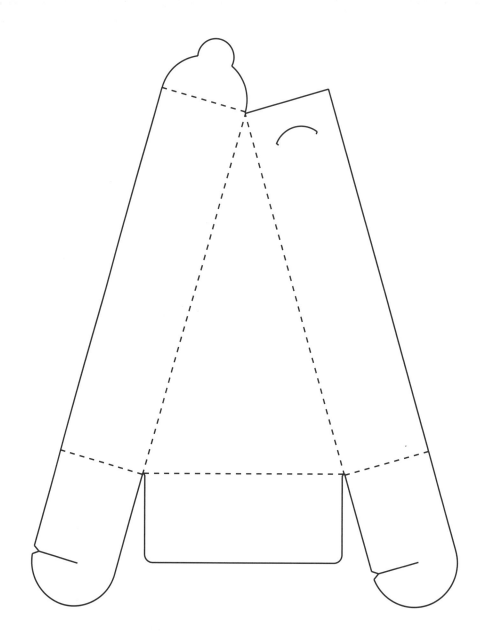

032

CHRISTMAS TREE PACK

Use: A surprise box with Christmas patterns. **Development:** Closed box consisting of two pieces that forms a rectangle secured with a sleeve. When the box is opened, we find two symmetrical pieces inside that, arranged vertically, assume the shape of a Christmas tree. Because of the die's shape, the sharp branches of the tree end with a hole in the lower part, making this box unsuitable for overly small objects, since they would fall out through the hole. **Material:** 280 g paperboard, 130 g tracing paper. → 1043

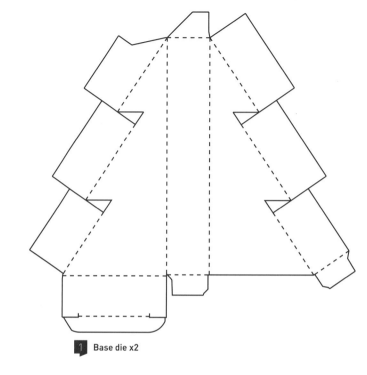

1 Base die x2

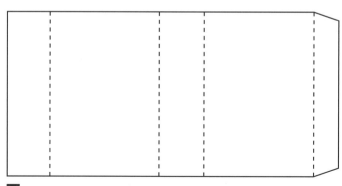

2 Sleeve die

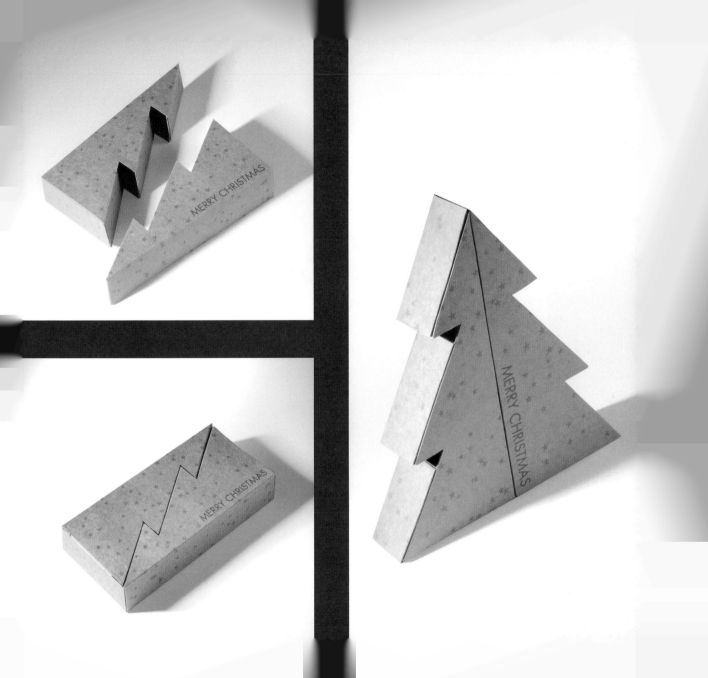

033

CARDBOARD HEART FOR BOTTLE

Use: To decorate wine bottles. **Development:** Simple heart-shaped piece. The inside is adapted to the outline of the bottleneck for a perfect fit. Small holes are added on the die where a bow will be placed. The bow will serve both as decoration and to hold an additional gift product in place. It is necessary to use double-layer corrugated cardboard as it possesses the rigidity required for this design. **Material:** 300 g paperboard, double-layer corrugated cardboard → 1848

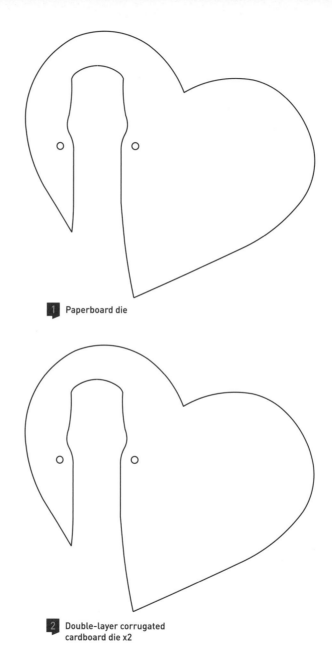

1 Paperboard die

2 Double-layer corrugated cardboard die x2

034

BALL INSERTED
IN A CROSS

Use: Simple design. Self-assembly. **Development:** The design consists of two square pieces that fit into each other and contain a partially visible golf ball. The ball is first introduced in the smaller box, and this small box into the larger one, like a sleeve. The two boxes are assembled with pressure to prevent the ball from falling. The result is a cross-shaped pack. **Material:** 300 g paperboard. → 1421

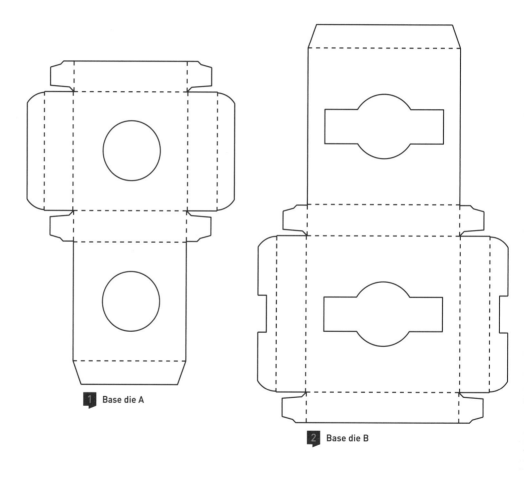

1 Base die A

2 Base die B

035

BOOSTER PACK

Use: Christmas box in the shape of a postcard for a personalized gift. **Development:** This box is perfect for photo gifts and is made of kraft cardstock with white print on the flap. **Material:** 300gr cardboard. → 2247

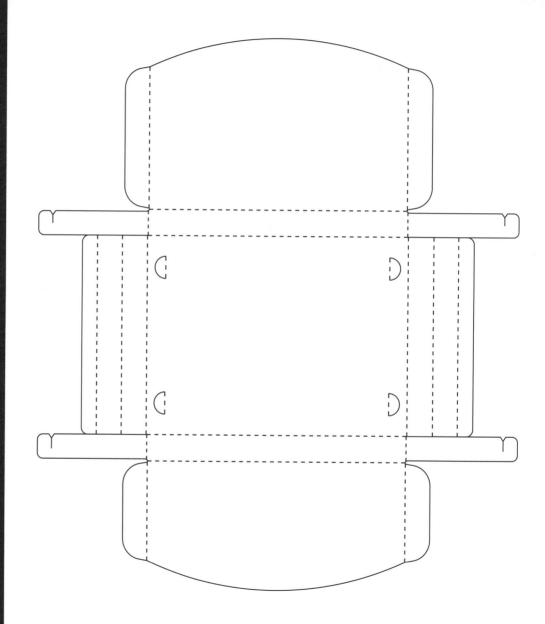

FOUR-BOTTLE STRUCTURE

Use: Simple packaging to carry and store four wine bottles. **Development:** A single die structure with a clever automatic system for opening and closing. The inside triangles at the top generate the upward motion to open the box and the downward motion to close it. Folding the leftover material of the windows protects the bottles during transport while simultaneously presenting the product. **Material:** Micro-flute cardboard. → 2035

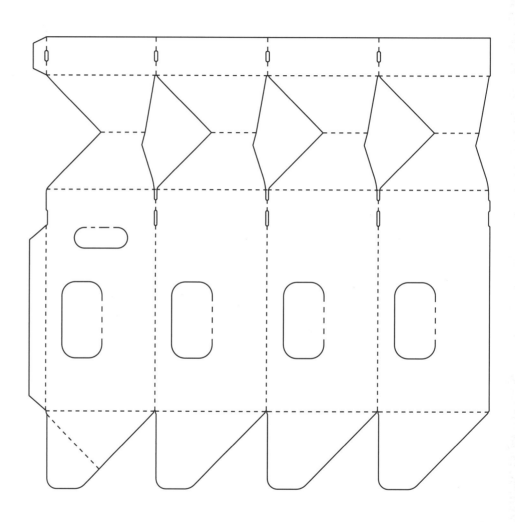

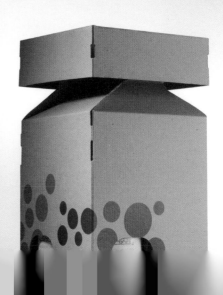

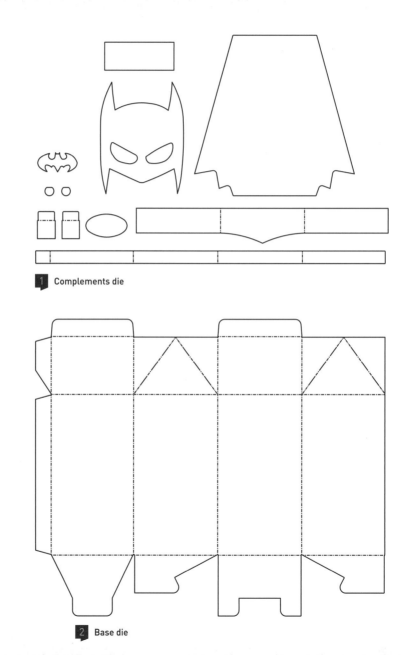

BATMAN
BOX

Use: The ideal box for a superhero-themed children's party. **Development:** Tetra-brick-shaped box that is combined with various die-cut parts of masks, belts, eyes and so on to create a superhero. **Materials:** 300 g cardboard. → 6003

1 Complements die

2 Base die

PERFUME
DIPTYCH

Use: Diptych for a perfume sample bottle. **Development:** In this single-die piece the perfume sample bottle is held in place through a slot. The diptych closes with a flap with an attractive sinuous curve. **Material:** 280 g paperboard. → 1050

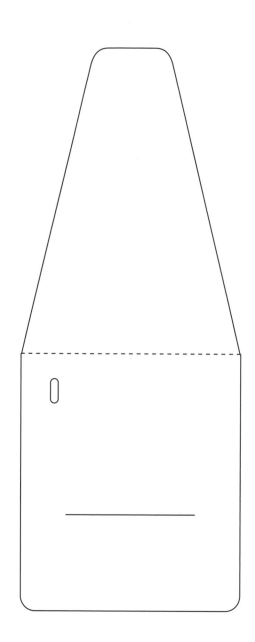

039

SURPRISE BOX
WITH "EARS"

Use: A simple easy to assemble box, ideal for desserts or small gifts. **Development:** This box's die is simple, making it easy to assemble. Notably, the two outside flaps reinforce the box and close it easily while at the same time giving it a fun look, as if they were ears. **Material:** 300 g paperboard. → 2214

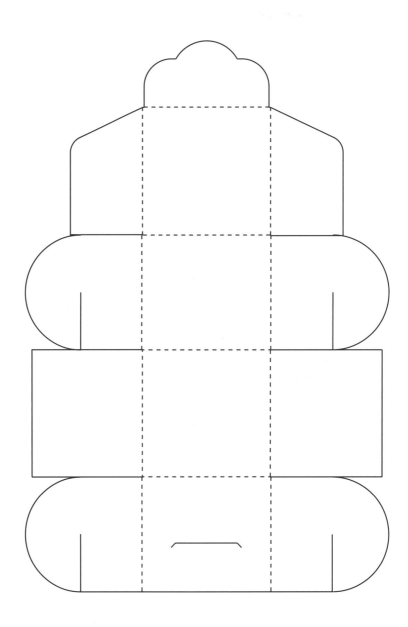

040

FOAM BOX
FOR USB

Use: Original box for small and delicate objects. **Development:** A single piece of EVA foam block, die-cut where the USB is going to go. Because it's a single block, its corners can be rounded. The paperboard triptych envelops the foam glued at the bottom. A small flap functions as a closing mechanism. Ideal for small formats. **Material:** 280 g paperboard, 15 mm EVA foam. → 2305

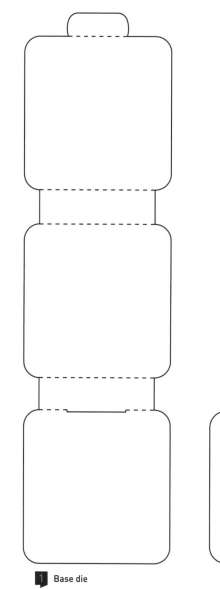

 Base die

 Foam die

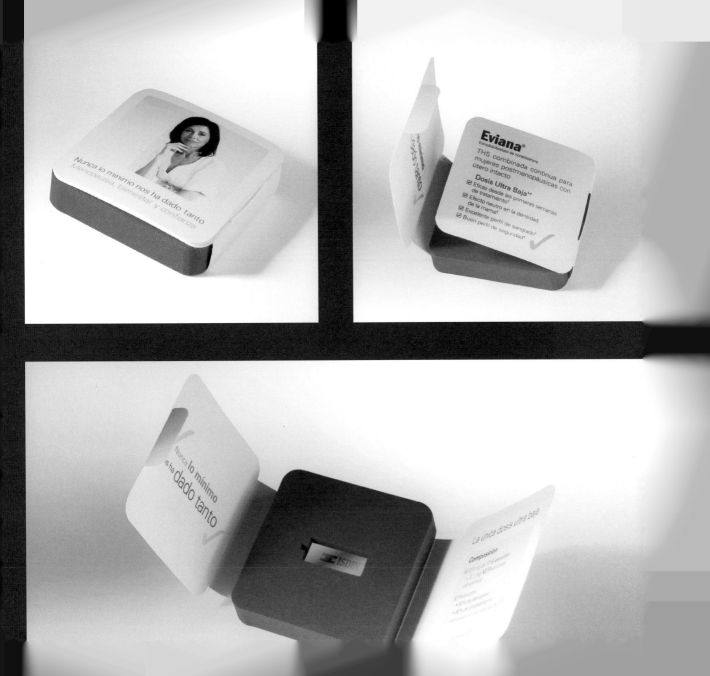

041

TRIANGLE WITH TRIANGULAR CLOSURE

Use: Simple box for bottles. **Development:** Self-assembly structure. No glue needed. It is recommended to use micro-flute cardboard to give the box the necessary strength. We add a velcro piece as well, glued to the rigid lid. At points of sales, these triangular boxes can be placed interspersed, thus optimizing the display space. **Material:** Micro-flute cardboard, 280 g paperboard, 18mm Ø velcro. → 1509

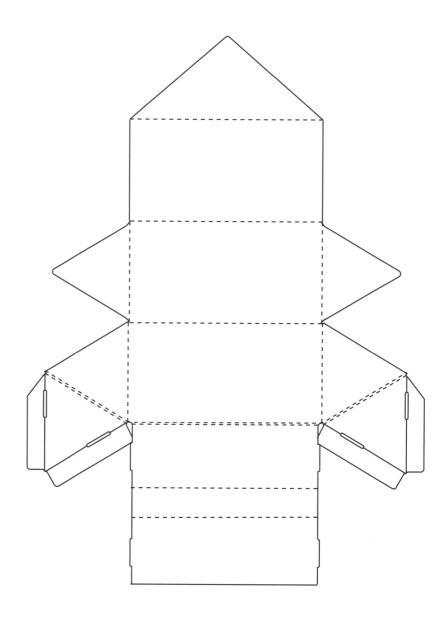

042

GIFT BOX
WITH CORD

Use: Simple construction. Self-assembled. **Development:** The base is made of gray cardboard for greater rigidity and better product support. Its triangular shape was created based on the same shape that appears in the brand's logotype. A standard folding stand holds the display by the back and keeps it at a certain rearward inclination. The visible borders of the cardboard should be kept in mind. **Material:** 300 g paperboard, 3 mm ø round tape. → 0280

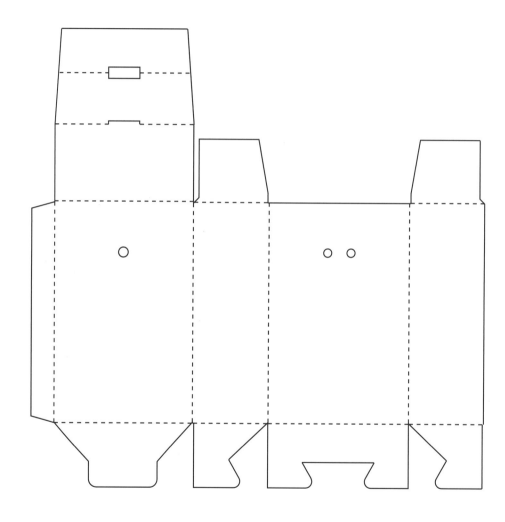

043

CAR-SHAPED BOX

Use: Simple structure car-shaped on the outside and that functions as a box. **Development:** This design makes the most of the structure in order to replicate a car in a very simple way. The center part is elongated and ends up being the box, while the sides acquire the "car" shape. These sides have the creases pushed inside so that when we assemble the packaging, the flaps fit in perfectly. **Material:** Micro-flute cardboard → 2360

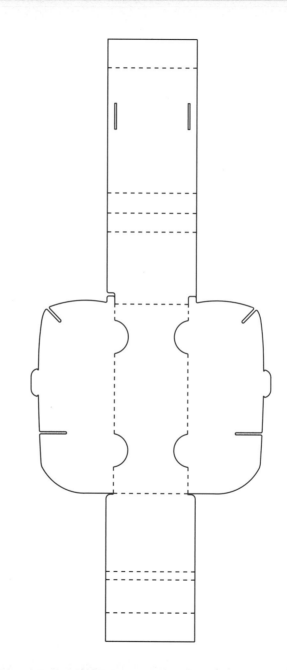

044

MATRYOSHKA BOX

Use: Simple die that can be elaborated in different sizes as a Matryoshka-like surprise packaging. **Development:** This cuboid box consists of a single die that can be folded and assembled without glue. A system of diagonal creases on the sides allows us to achieve stronger box. When opened, the box's folds exert pressure outward, producing a surprise effect. **Material:** 300 g paperboard. →1506

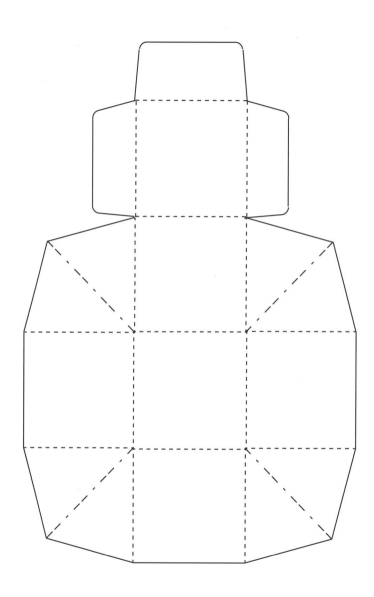

045

TWO EMBEDDED TRIANGLES

Use: Two embedded boxes for small objects. **Development:** This set consists of two embedded triangle-shaped boxes of the same size. One of them serves as a clamp, closing the other triangle. Both boxes can contain objects. The die needs to be very precise so that both triangles fit perfectly. **Material:** 300 g paperboard. → 1422

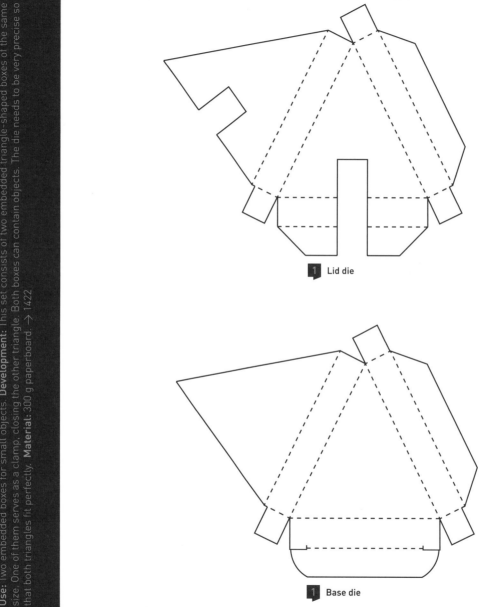

1 Lid die

1 Base die

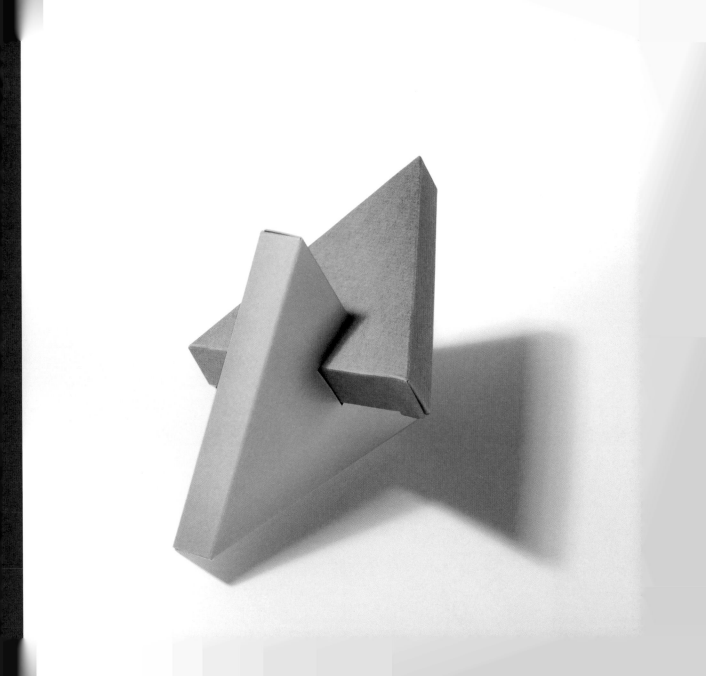

046

Use: Small-format box with fastening tabs. **Development:** This box is ideal for small and precious gifts. Printed in cold tones, its structure allows the design to be customized. **Material:** 300gr cardboard. → 6065

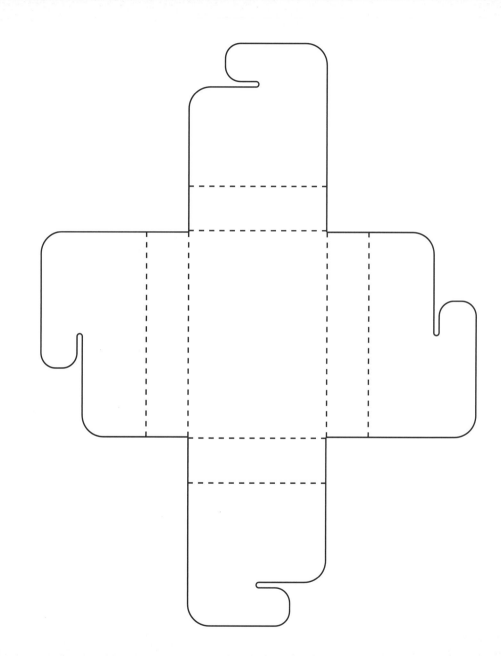

TIED-TOGETHER TRIANGLES

Use: Welcome pack for hotel guests. **Development:** Three triangle-shaped boxes consisting of a simple die. The two lateral boxes are the same size while the middle one is larger. They're tied together with a string passed through a hole on top of the triangles, thus becoming a single pack. If used for food products, the inside varnish needs to be suitable for this type of application. **Material:** 300 g paperboard, 3 mm ø string and a clip. → 1423

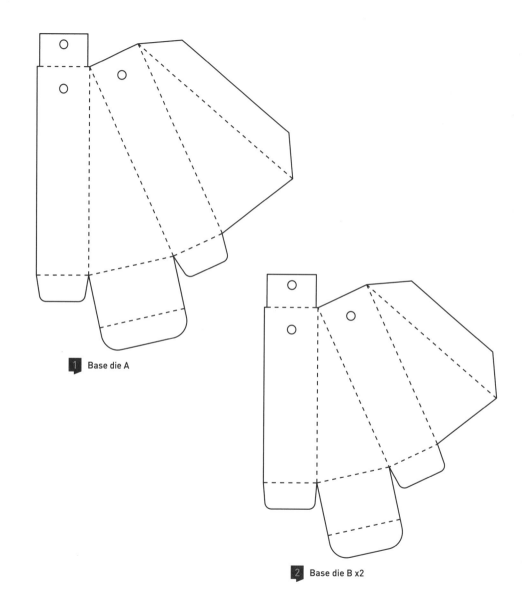

1 Base die A

2 Base die B x2

048

GIFT BAG
WITH HANDLE

Use: Rigid bag for small objects, ideal for shops. **Development:** This auto bottom bag is based on a simple structure glued by one of its sides. The top closes with a flap and includes a hole that functions as a handle, making the pack easy to carry. This structure is only suitable for small formats. **Material:** 300 g paperboard. → 2001

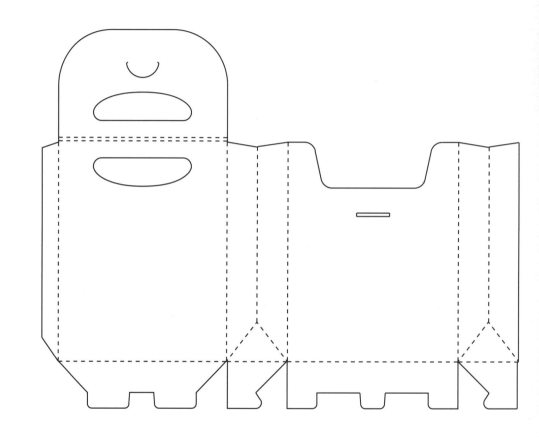

/224/ /116/ /232/ /124/ /126/ /128/ /246/

/140/ /172/ /236/ /240/ /160/ /156/ /202/

/178/ /164/ /150/ /184/ /192/ /216/ /228/

/132/ /204/ /144/ /212/ /168/ /220/ /112/

/244/ /182/ /188/ /198/ /120/ /146/ /196/

/136/

/138/

/152/

/174/

/208/

/LEVEL02/

"TWISTED" BOX FOR BOTTLES

Use: Unconventional gift box for bottles. **Development:** Several overlapping layers of corrugated cardboard make up this box, ideal for giving bottles as a gift. The layers are held together with a fabric ribbon affixed inside in the first and last pieces. The box therefore can be shaped by hand, experimenting with different forms without sacrificing the vertical structure of the packaging. Placement of an exterior sleeve is recommended to secure the pack. **Material:** 300 g paperboard, double-layer corrugated cardboard, 18 mm wide ribbon. → 1212

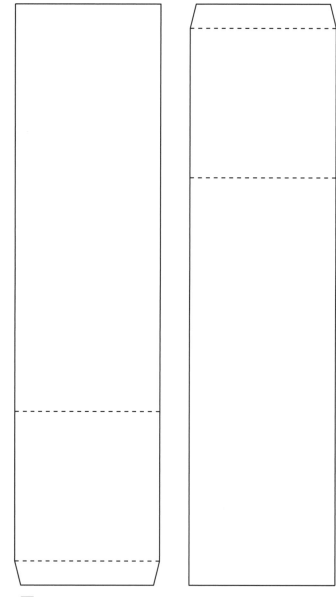

 Sleeve die

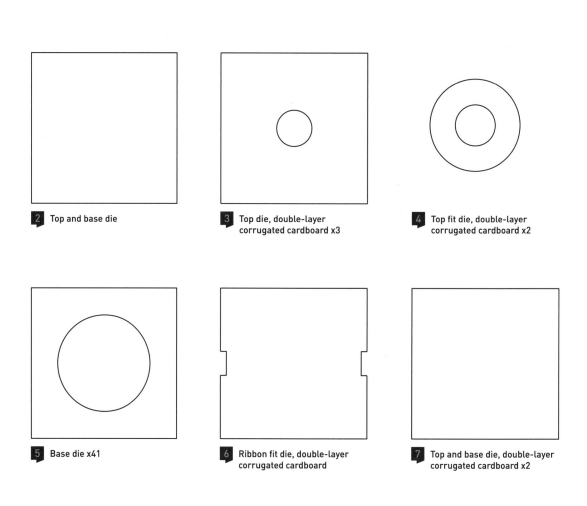

2 Top and base die

3 Top die, double-layer
   corrugated cardboard x3

4 Top fit die, double-layer
   corrugated cardboard x2

5 Base die x41

6 Ribbon fit die, double-layer
   corrugated cardboard

7 Top and base die, double-layer
   corrugated cardboard x2

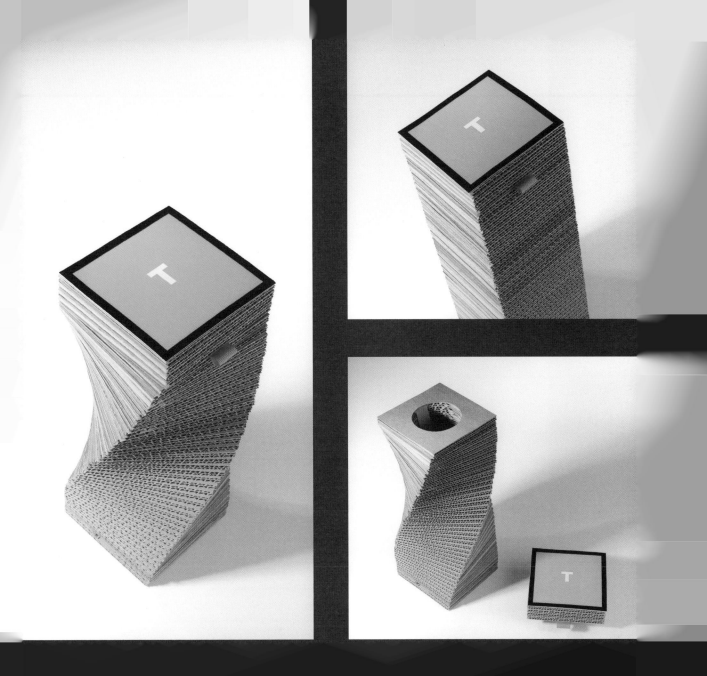

1 Base die A

product and keep it visible. When we separate each part, the box acts as a product dispenser and facilitates extraction. The box is closed with a transparent sleeve that hints at what is inside. **Material:** 300 g paperboard, micro-flute cardboard, 5 mm foam, and 130 g tracing paper. → 2044

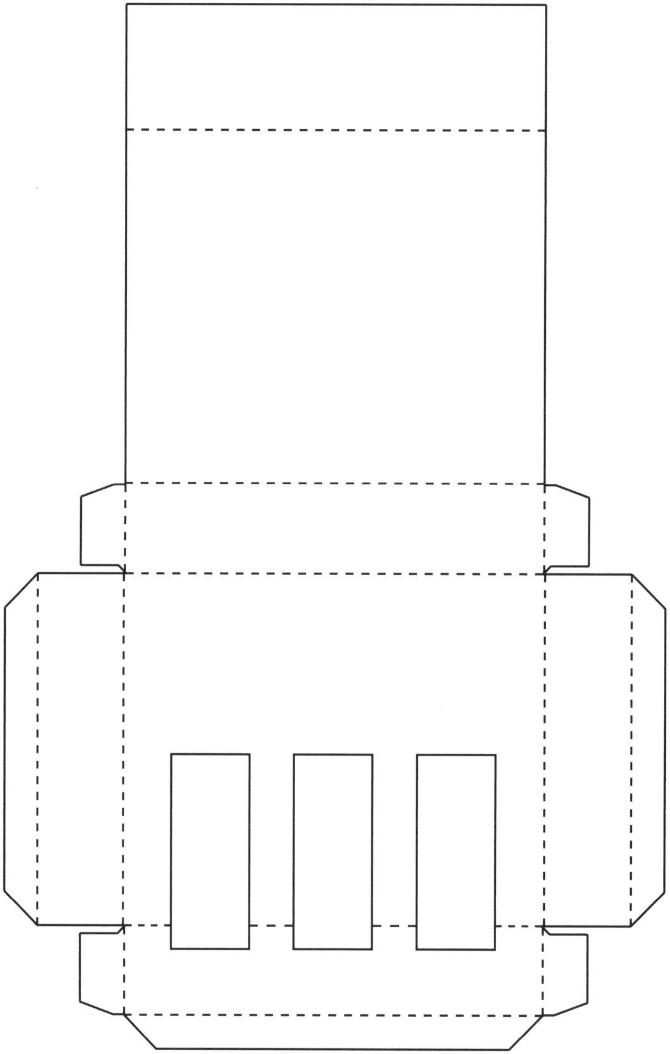

LA VUELTA AL MUNDO EN TRES SABORES

Después de más de 25 años, nuestra principal inquietud sigue siendo hacer que nuestros clientes se sientan bien acogidos en nuestros hoteles.

Por eso en H10 Hotels le ofrecemos cada día más destinos donde descansar, trabajar y divertirse sintiéndose como en casa, preocupándonos tan sólo de disfrutar de diferentes tradiciones y culturas.

Una buena manera de empezar a conocerlas es a través de esta colección de chocolates que hemos seleccionado para usted y que le invitamos a degustar.

Felices Fiestas

ALEMANIA
Desde la factoría de Germania surgen una larga creatividad de formas de elaborar chocolate que se intensifica expresando en su sabiduría con la refinación para los paladares más exigentes.

32%

ESPAÑA
Los mejores nuestros chocolates españoles, con renovada renacen, transcender aprovechar la tradición, unidos de nuevos gustos propuestas creativos con marcos prensados y experiencia centenaria.

50%

MÉXICO
El descubrimiento del Nuevo Mundo llevó consigo años atrás, descubriendo el cacao, a partir del cual se elaboran los más exquisitos dulces y chocolates que se consumen alrededor de todo el mundo.

70%

H|10 HOTELS
Pensando en ti

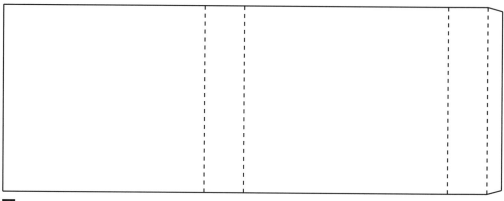

2 Sleeve die

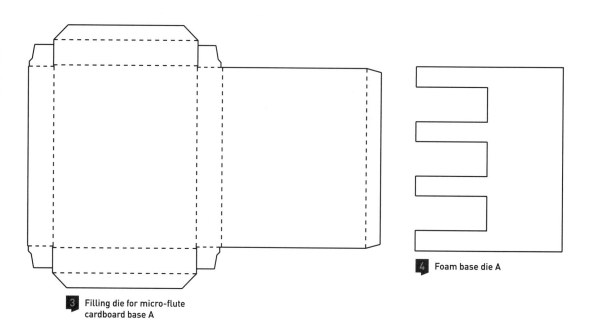

3 Filling die for micro-flute cardboard base A

4 Foam base die A

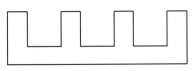

5 Foam base die B

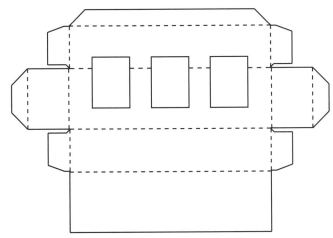

6 Base die B

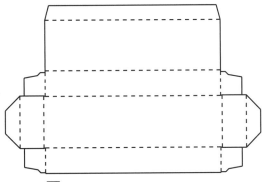

7 Filling die for micro-flute cardboard base B

051

SOFTWARE TRIPTYCH

Use: A box for software that includes a CD and manuals. **Development:** The box opens like a book. A cardboard hinge connects the interior trays, thereby distributing the content by arranging the different elements: CDs, manuals, warranties... The exterior piece doesn't cover the whole box, offering a playful visual composition of both pieces. **Material:** 300 g paperboard, 5 mm foam, double-layer corrugated cardboard, grey cardboard. → 2178

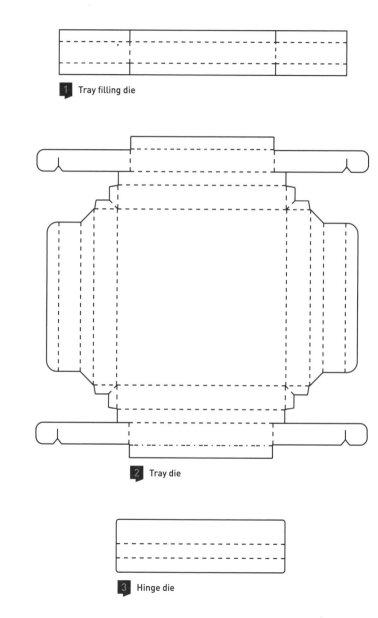

1 Tray filling die

2 Tray die

3 Hinge die

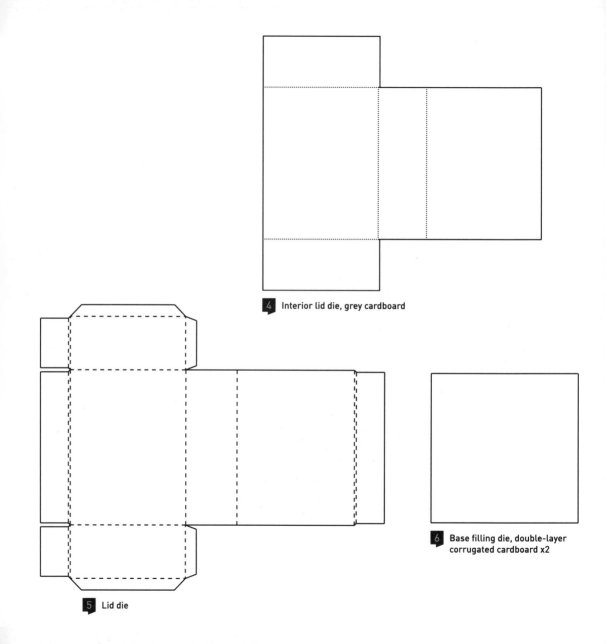

4 Interior lid die, grey cardboard

6 Base filling die, double-layer corrugated cardboard x2

5 Lid die

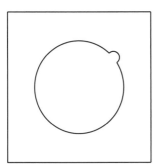

7 Base foam filling die

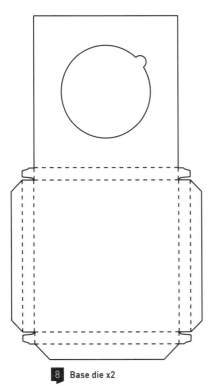

8 Base die x2

052

DOUBLE-LAYER CORRUGATED CARDBOARD SLEEVE

Use: Box that functions as a cover for delivering a book. **Development:** This pack consists of several layers of perforated double-layer corrugated cardboard that are glued together to form a sleeve that acts as a creative cover for a book. The upper part includes a rectangular cavity where a corporate name or logo is placed. The rounded edges of the pieces of double-layer corrugated cardboard result in a more resistant pack not easily damaged by impacts. **Materials:** 300 g paperboard, double-layer corrugated cardboard →

2369

1 Central pieces die, double-layer corrugated cardboard x39

2 Exterior pieces die x12

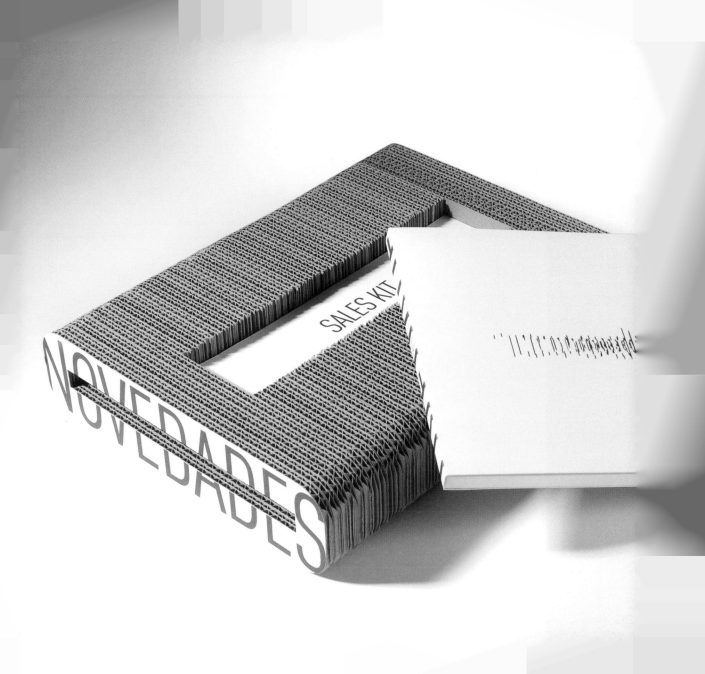

053

CYLINDER FOR CREDIT CARDS

Use: This tube box has a simple, original design. Ideal for delivering credit cards. **Development:** It consists of two simple dies that fold into a tube and create the two parts of the box. The interior piece includes a groove for placing the card. The box is closed by inserting the cylinder containing the card inside the other cylinder while keeping the card visible from the sides. **Material:** 300 g paperboard. Box designed by ELISAVA (Beth Puigbó) → 1765

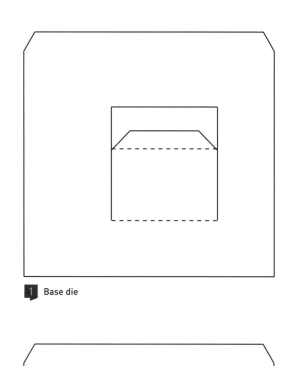

1 Base die

 2 Sleeve die

054

FOAM
TRIPTYCH BOX

Use: A box with a central piece that holds the product. **Development:** The box consists of two trays and a central piece in which the product is embedded. This foam piece is die-cut in such a way that it adapts to the product and holds it tight. At the same time, part of the product juts out. The box is closed with elastic bands held together with a piece of cardboard. **Material:** 280 g paperboard, 20 mm foam, micro-flute cardboard, 2 mm grey cardboard and 2 mm Ø elastic band. → 1740

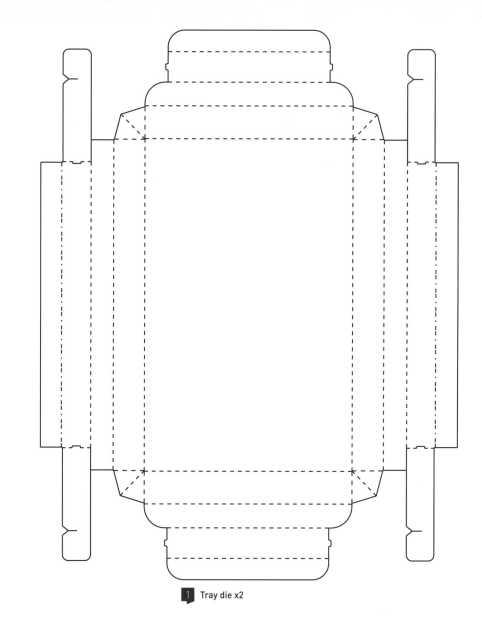

1 Tray die x2

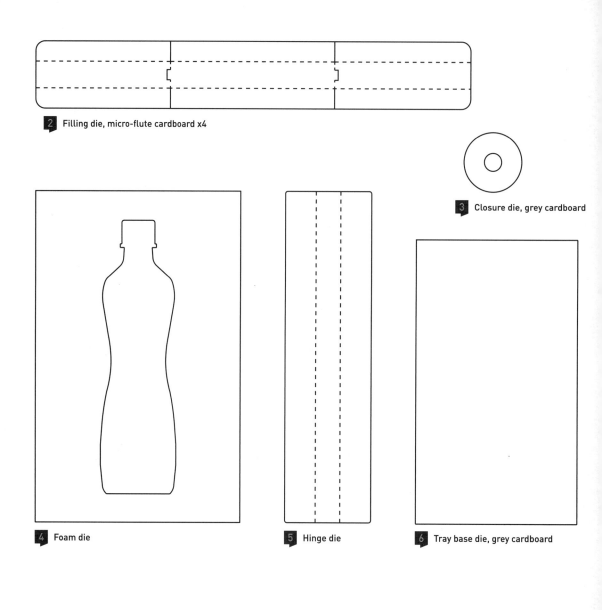

2   Filling die, micro-flute cardboard x4

3   Closure die, grey cardboard

4   Foam die

5   Hinge die

6   Tray base die, grey cardboard

Use: Small box for cosmetics with an unconventional opening. **Development:** Simple box consisting of a base and two sleeves that fit one into the other to close the pack. The base includes a space where the product is placed, creating a surprise effect when the pack is opened. The height of the sleeve must be adjusted to the dimensions of the object to prevent the latter from moving when closing the package. **Materials:** 300 g paperboard. → 2361

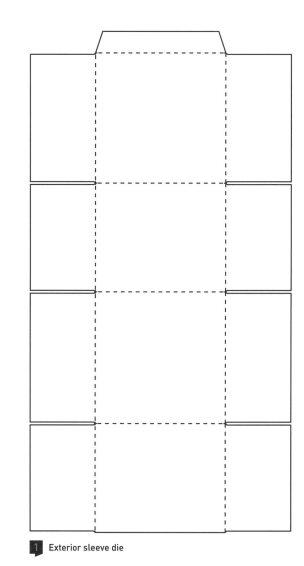

1 Exterior sleeve die

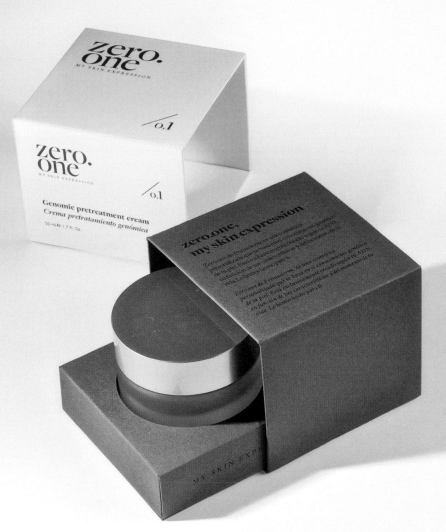

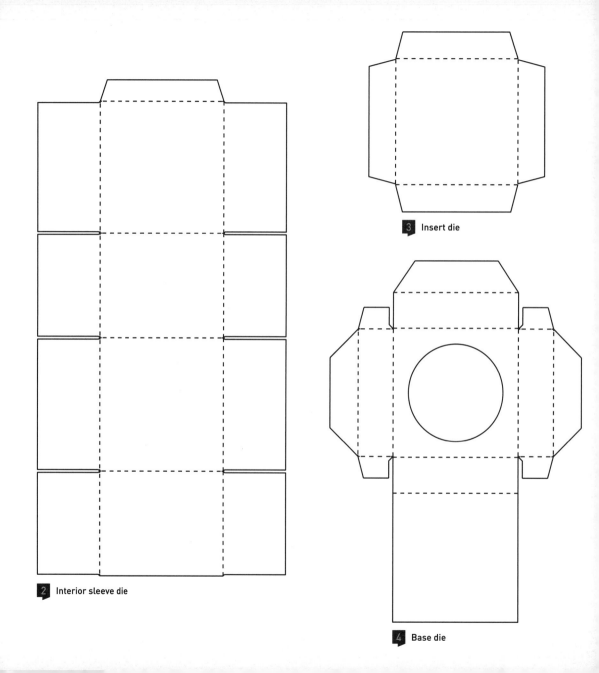

2 Interior sleeve die

3 Insert die

4 Base die

056

CURVED PLASTIC PACKAGING

Use: A die-cut display made of plastic to showcase a product sample. **Development:** An original way to showcase a product, given that this packaging consists of two plastic pieces connected with two rivets. The product, a switch in this case, is anchored in the interior plastic piece, becoming part of the packaging. It is closed with another external rivet. The product is visible thanks to the see-through material. **Material:** 0.3 mm PET plastic, rivets and snap fastener. → 1037

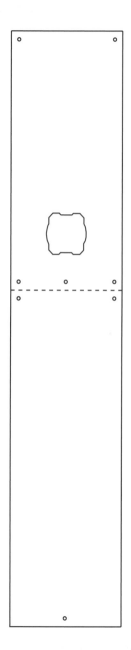

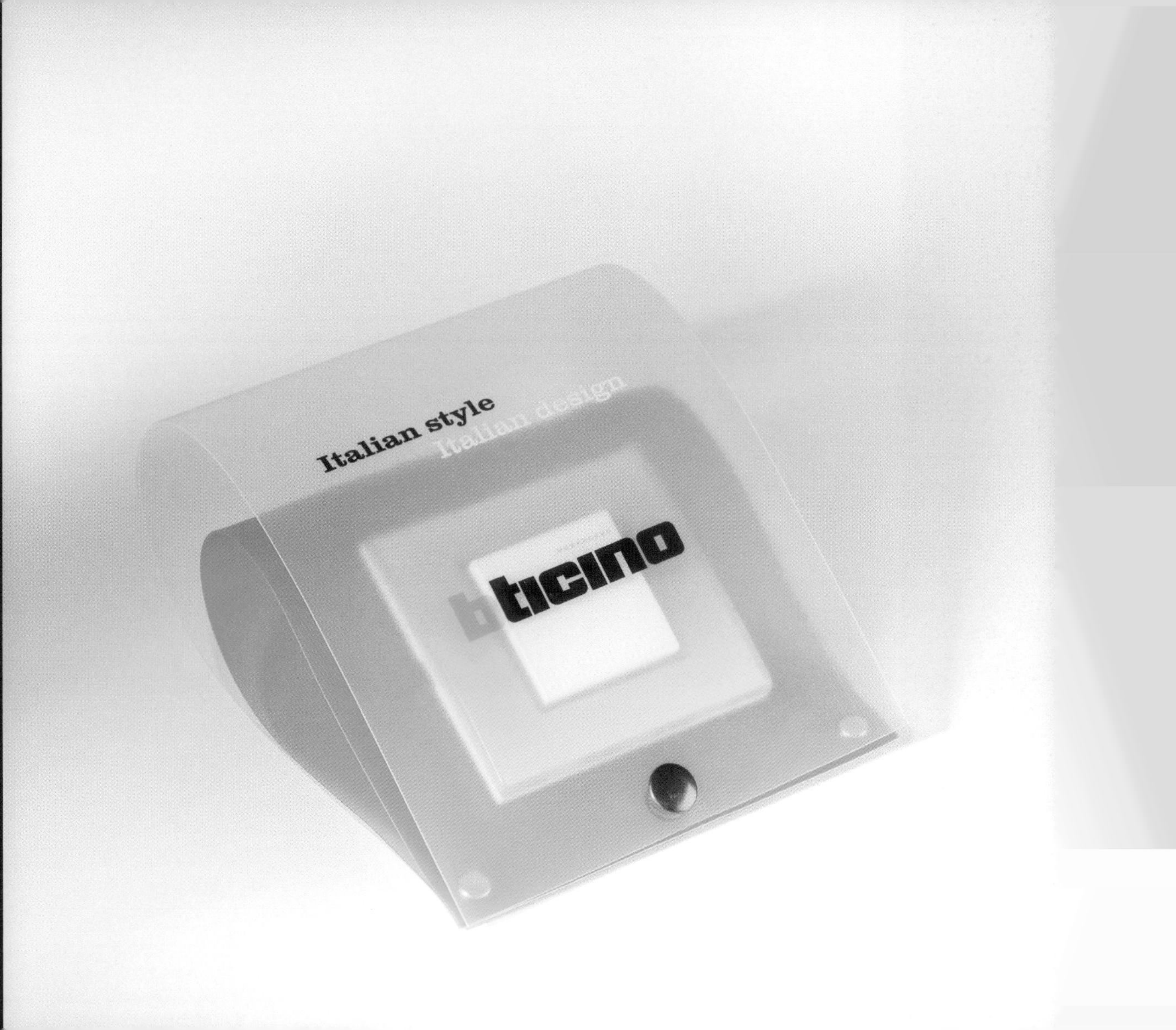

Italian style

Italian design

bticino

057

SIX TRIANGLE CUBE

Use: A cube with six boxes for an assortment of products. **Development:** Packaging consisting of six triangular boxes. When assembled, the result is a cube held together by a decorative string. We can decorate the exterior by adding labels and using the string in a playful manner. If used for food products, the inside varnish needs to be adequate. **Material:** 280 g paperboard. 3mm ø string. → 2306

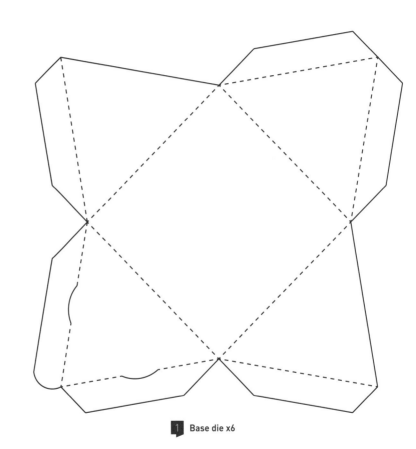

1 Base die x6

MULTI-LAYER
SAMPLE BOX

Use: For product samples. **Development:** Superimposing several pieces of double-layer corrugated cardboard creates the base and lid. The shape of the product is die-cut to keep it snugly held in place while simultaneously allowing it to jut out a bit. This box stands out for combining noble materials such as printed fabric with rustic ones such as the corrugated cardboard. **Material:** Double-layer corrugated cardboard, 18 mm Ø magnet, canvas paper, 300 g paperboard and 3 mm Ø elastic band. → 2041

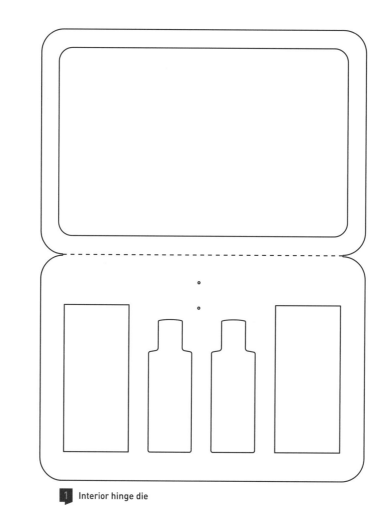

1 Interior hinge die

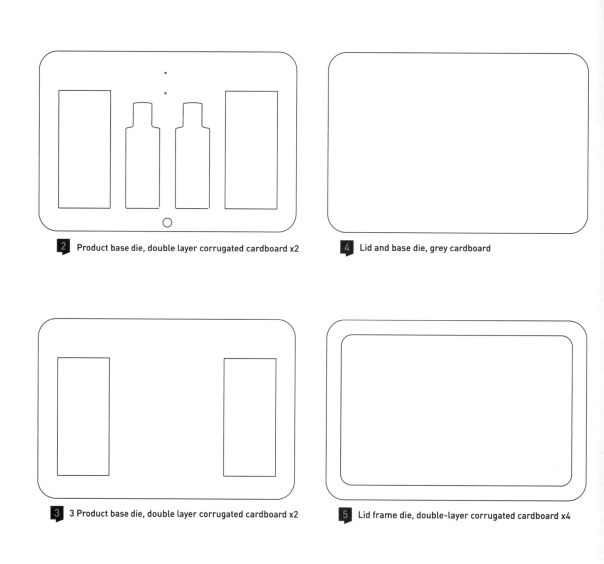

2 Product base die, double layer corrugated cardboard x2

4 Lid and base die, grey cardboard

3 3 Product base die, double layer corrugated cardboard x2

5 Lid frame die, double-layer corrugated cardboard x4

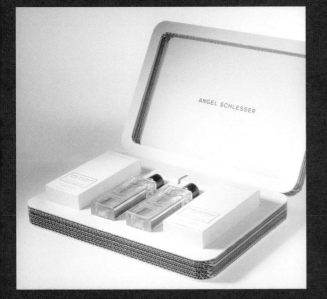

059

TWO INTERWEAVED TRIANGLES

Use: An original star-shaped box, ideal for chocolates or small gifts. **Development:** This box stands out because of its 3D star shape. The structure consists of two pieces with grooves and holes that allow one part to fit into the other when assembling the pack. The box includes a flap that helps open it without having to disassemble the pack. **Material:** 300 g paperboard. → 2036

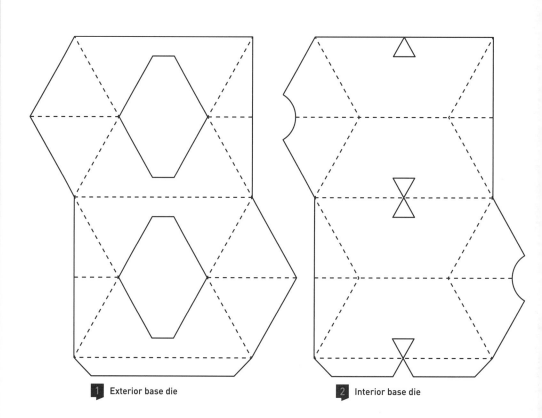

1 Exterior base die

2 Interior base die

060

TWO-COLOR SURPRISE BOX

Use: This box's opening mechanism creates a sensation of wonder and highlights its content. **Development:** It consists of a lid and base made of a single piece than can be self-assembled. From the outside, it looks like a regular bottle box, yet when the lid is lifted, the folds open showing and highlighting the product. When it is closed, these folds make the box stronger and hold the product snugly in place. **Material:** 350 g paperboard → 2002

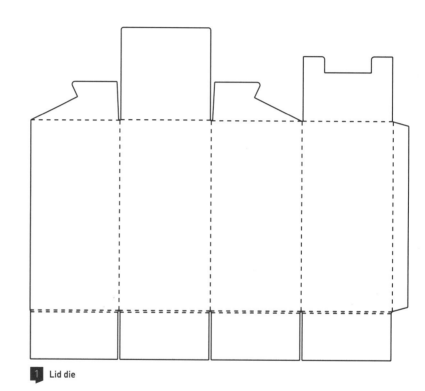

1 Lid die

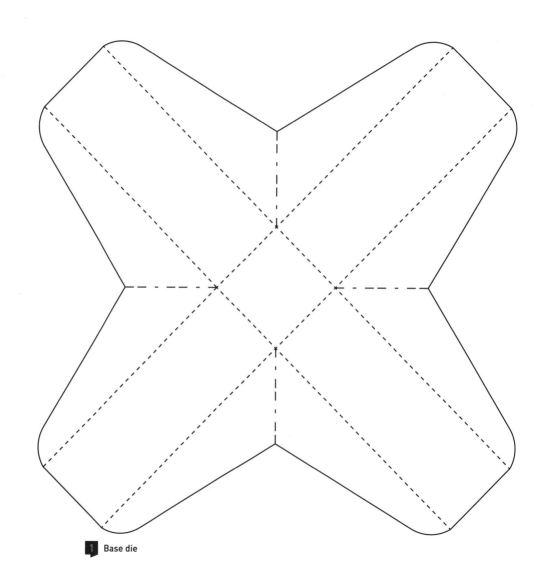

1 Base die

061

BLOCK FOR CD

Use:: A block for a CD that highlights the product. **Development::** By overlapping multiple pieces of double-layer corrugated cardboard, this pack conceals a CD in a wide box that highlights the product. It consists of a printed lid and a base for the product, which is only half-shown. It closes with magnets hidden behind the paperboard that covers the interior of both the lid and the base. **Material::** Double-layer corrugated cardboard, 10 mm Ø magnet and 250 g paperboard. → 0548

1 Lid and base die

2 Lid and base interior die x14

3 Lid and base magnet die

062

TWO
TRAY PACK

Use: Display box to highlight a book. **Development:** Simple box that consists of two separate trays. One has an interior piece that sticks out and holds the book, which is highlighted as a result when opening the box. The other tray serves as a lid, adjusting snugly thanks to the interior piece that also makes room for the book. **Material:** 280 g paperboard, micro-flute cardboard, 18 mm Ø magnet and strapping band. → 1001

1 Tray die x2

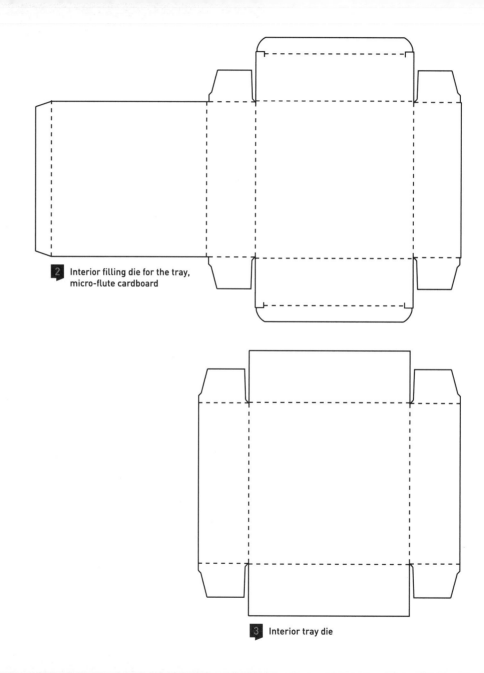

2 Interior filling die for the tray,
micro-flute cardboard

3 Interior tray die

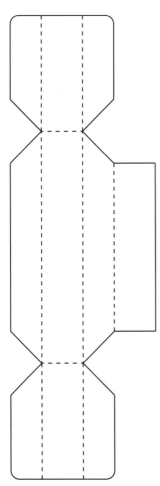

4 Interior filling die for the tray, micro-flute cardboard x4

063

BOOK
WITH GRASS

Use: A welcome box with synthetic grass on top. **Development:** This pack made with double-layer corrugated cardboard includes synthetic grass on top. The idea is to communicate the concept of sustainability explained in the informative brochure located inside. The back has a hole for a memory card with product information. Simple materials are used to evoke nature. **Material:** 300 g paperboard, double-layer corrugated cardboard, synthetic grass. → 2038

1 Cover die, double-layer corrugated cardboard

2 Base die, double-layer corrugated cardboard

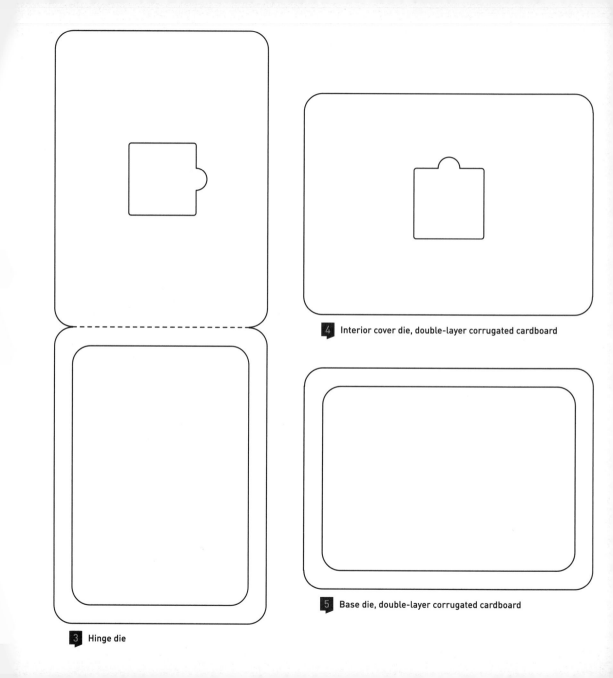

4 Interior cover die, double-layer corrugated cardboard

5 Base die, double-layer corrugated cardboard

3 Hinge die

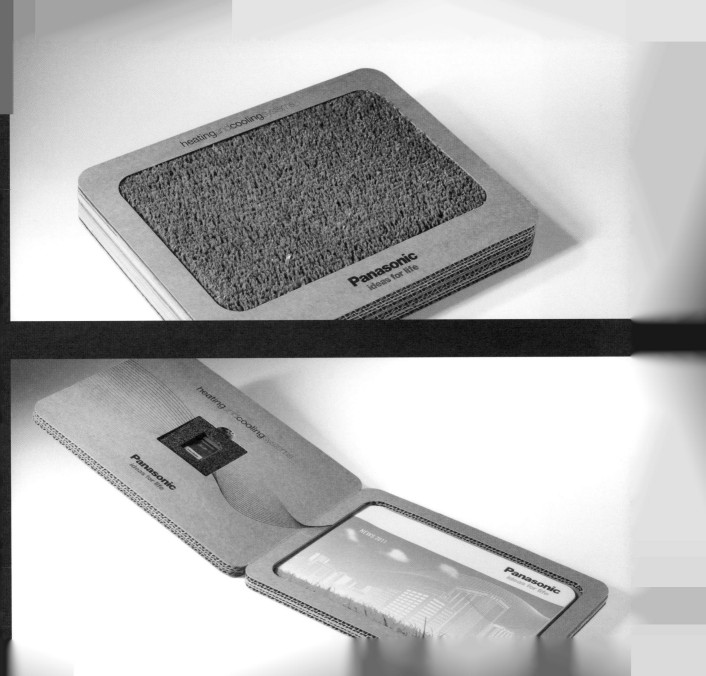

064

MULTI-USE
BOX-BAG

Use: A bag that incorporates a box to carry products. **Development:** A bag made with EVA foam that contains a cardboard box where the product goes. The bag has holes that function as handles, and the logo is die-cut in the foam, revealing the color of the box inside. **Material:** 300 g paperboard, 2 mm EVA foam. → 1336

nu

INTIM

 1 Hinge die

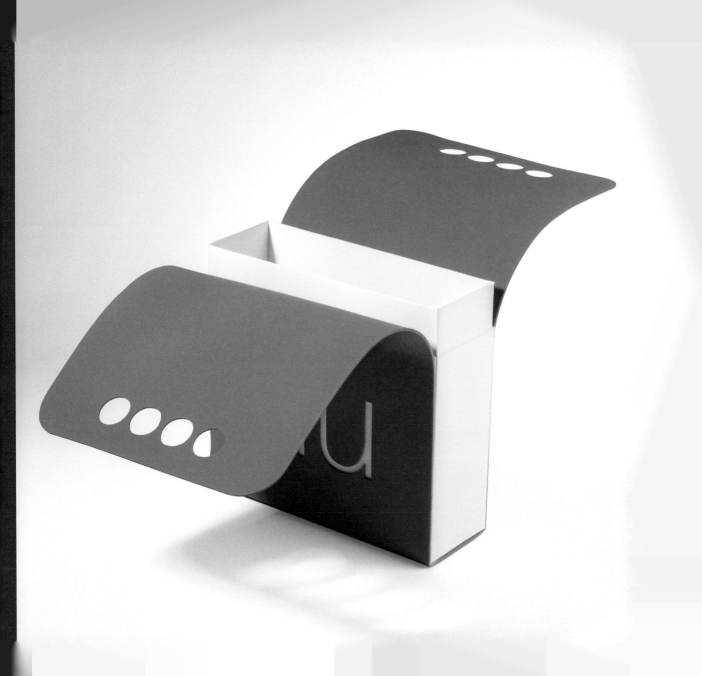

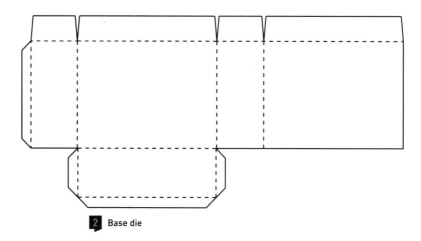

2 Base die

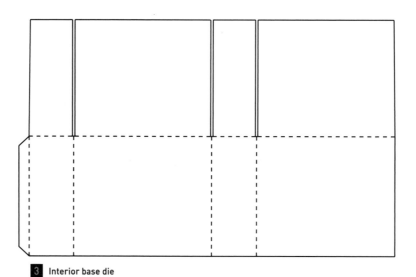

3 Interior base die

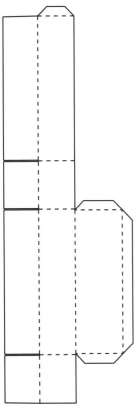

 Lid die

065

MULTI-LAYER CD TRIPTYCH

Use: Pack consisting of three pieces that open like a book. **Development:** The packaging consists of three pieces: two function as a cover and back cover while the central piece is the hinge. The covers are made with double-layer corrugated cardboard, presenting a suggestive combination of materials. The pack is closed with a sleeve. **Material:** 300 g paperboard, double-layer corrugated cardboard. → 1855

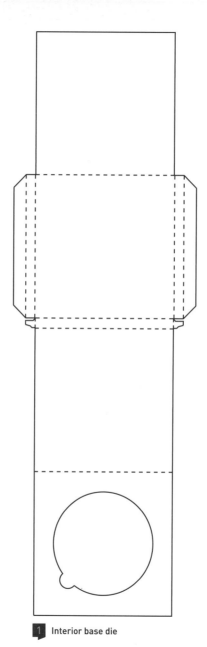

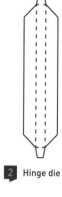

2 Hinge die

1 Interior base die

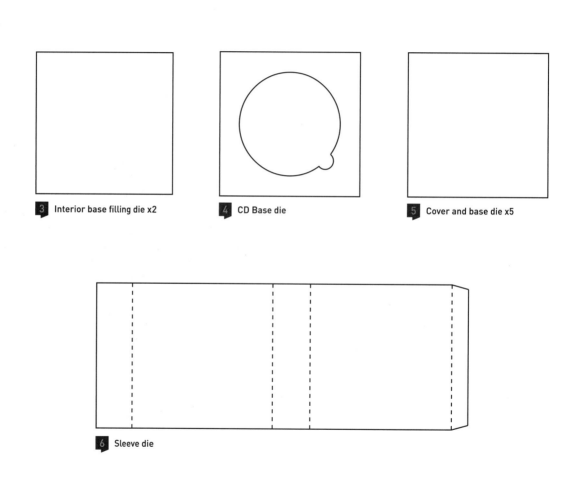

3 Interior base filling die x2

4 CD Base die

5 Cover and base die x5

6 Sleeve die

066

TRIPLE
DISPLAY BOX

Use: A welcome box that functions as display as well. **Development:** This packaging consists of three trays and a removable interior foam piece to which the product is affixed. If we remove this foam piece, we can set it vertically on the lid's gap, transforming the welcome box into a display. **Material:** 300 g paperboard, 12 and 45 mm foam. → 0996

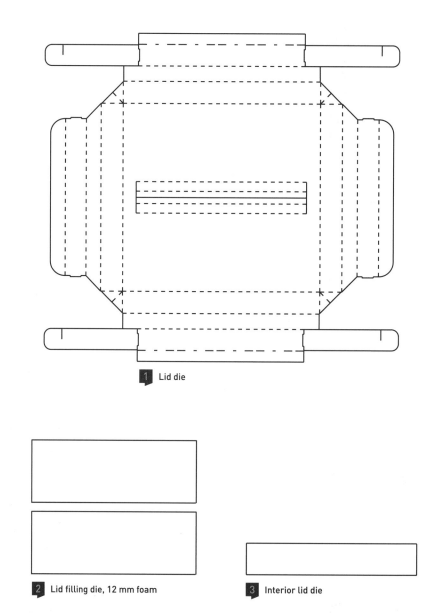

1 Lid die

2 Lid filling die, 12 mm foam

3 Interior lid die

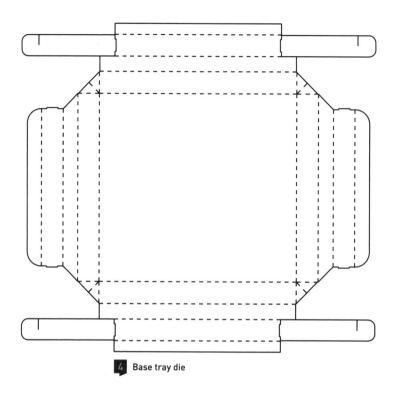

4 Base tray die

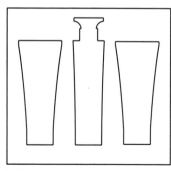

5 45 mm foam die

6 Interior tray die x2

067

NOTE
PAD FOLDER

Use: A cleverly designed rigid folder with a note pad, ideal for presentations. **Development:** Made with grey cardboard glued to paperboard, this folder has small orifices where the string's knots are located to close the folder. The pad is attached to the die-cut circle glued to the folder's inside flap. Keep in mind the thickness of the pad when calculating the exterior spine. **Material:** 2 mm grey cardboard, 200 g paperboard, 3 mm Ø elastic band. → 2014

1 Folder die

2 Pad cover die

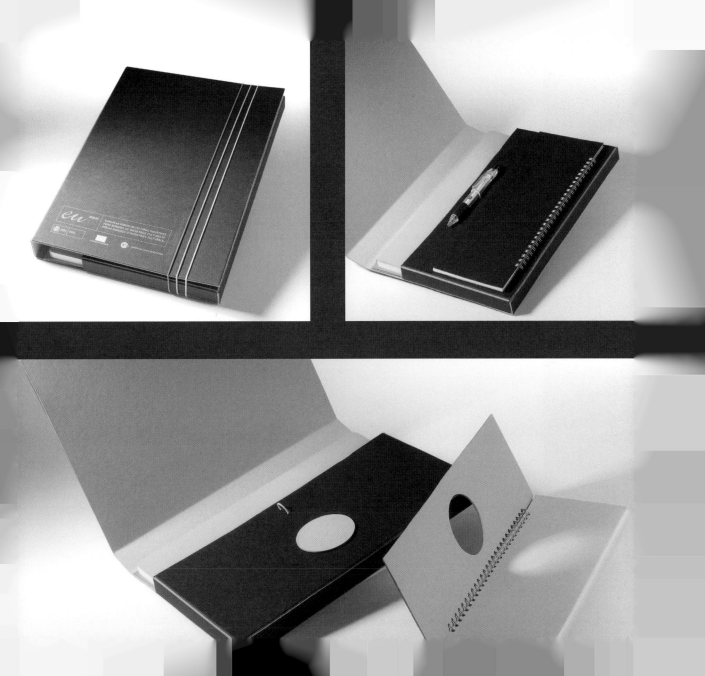

068

GRASS
BETWEEN BLOCKS

Use: An attractive box with artificial grass that grants it an air of originality. **Development:** The middle part, made of artificial grass and the place where event brochures go, is what makes this packaging unique and different. It has a cardboard base and a lid in the shape of a tray with wide walls, adding symmetry and volume to the box. To save costs, the lid and base are made with the same die. **Material:** 300 g paperboard, double-layer corrugated cardboard and artificial grass. → 2156

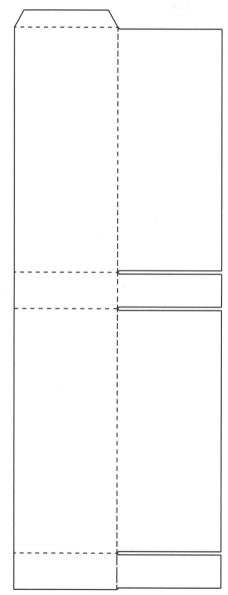

 Interior base die

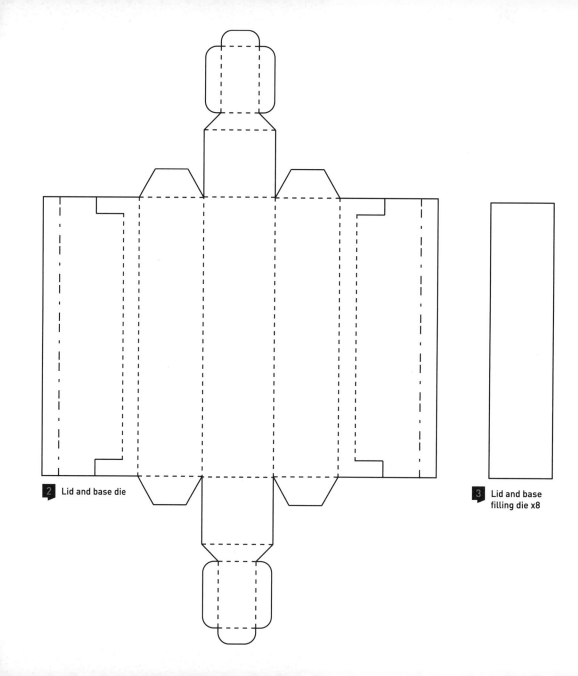

2 Lid and base die

3 Lid and base filling die x8

4 Artificial grass die

069

THREE
LAYER PACK

Use: The idea is to create a surprise effect when opening this apparently simple box. **Development:** The recipient is surprised by the way the box opens like a fan. The three trays are based on the same die, along with another for the sides to which the trays are glued. Depending on the content, the middle tray can be reinforced with another lining going all the way to the bottom. The lid helps keep all the trays together. **Material:** 280 g paperboard → 0321

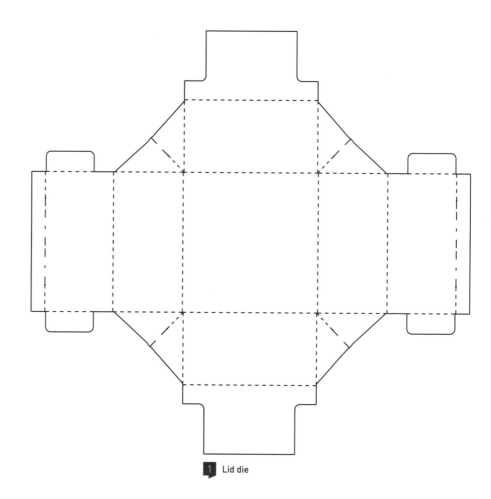

1 Lid die

SÓLOCACAO

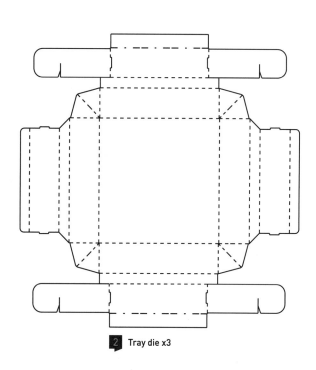

2 Tray die x3

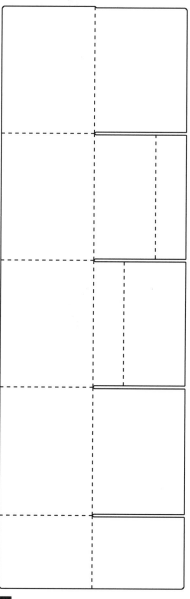

3 Hinge die

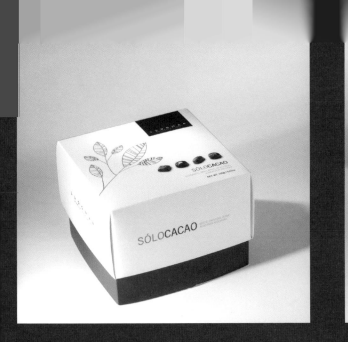

070

GIFT BOX FOR BOTTLES

Use: Gift box for bottles, with lid and string for easy transport. **Development:** This box for bottles is based on a simple die and has a piece inside that reinforces its structure. The string gives it a touch of originality in addition to lifting the lid to close the box.
Material: 2 mm grey cardboard, 250 g paperboard, 8 mm Ø string, 8 mm Ø string. → 1502

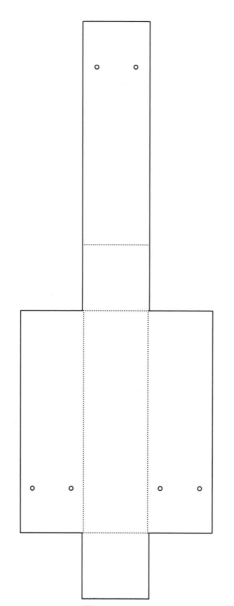

1 **Exterior die**

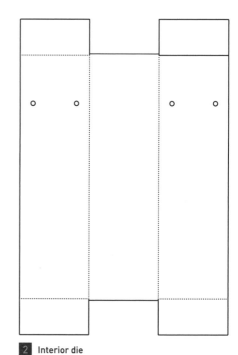

2 **Interior die**

071

BOX WITH RAISED BASE

Use: A box with a lid that turns into two pillars to display the product. **Development:** This box consists of a tray base where the product is placed and two lids that, when opened, can be set as pillars to highlight the product and raise it. Keep in mind that a flap prevents the two lids from ever separating from the base. **Material:** 300 g paperboard. → 2016

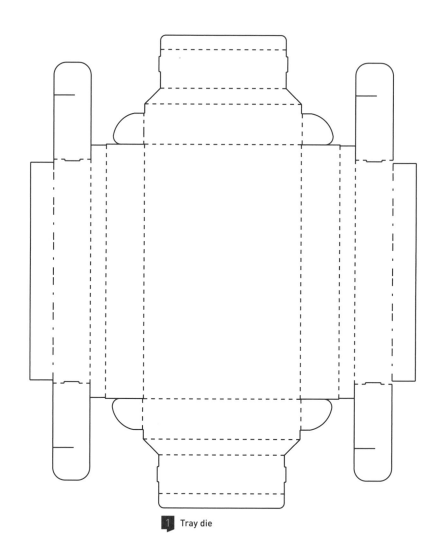

1 Tray die

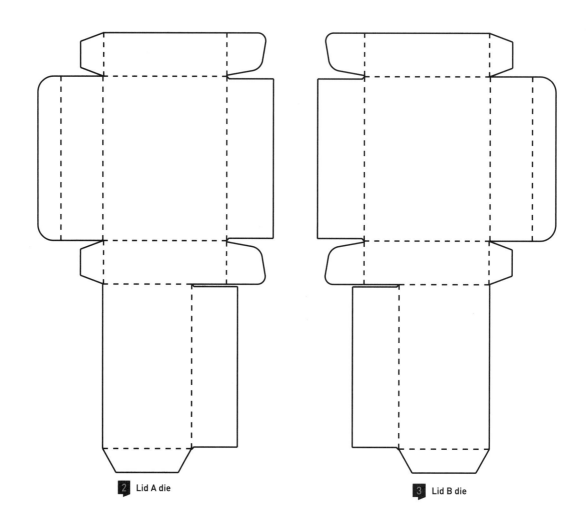

2 Lid A die

3 Lid B die

072

GIFT CASE WITH SLEEVE

Use: A welcome box for a perfume launch. **Development:** This display box includes a thick sleeve made of cardboard that draws inspiration from the perfume's packaging design. The die consists of a tray for the lid and a base with grooves that helps highlight the product. **Material:** 300 g paperboard, double-layer corrugated cardboard, 20 mm foam, two 18 mm Ø magnets. → 2049

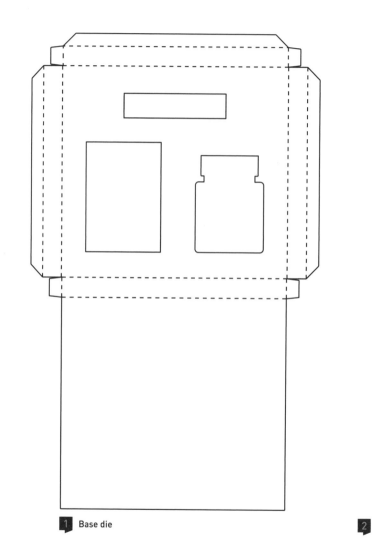

1 Base die

2 Hinge die

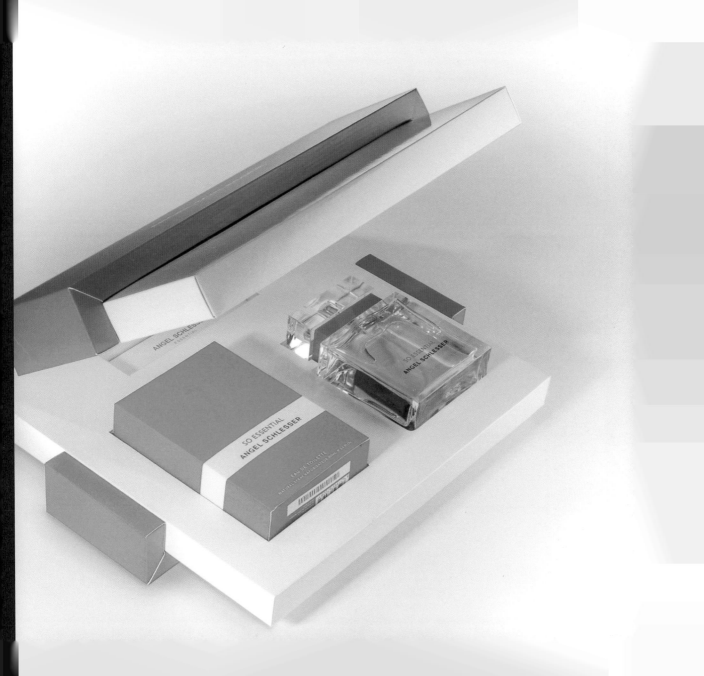

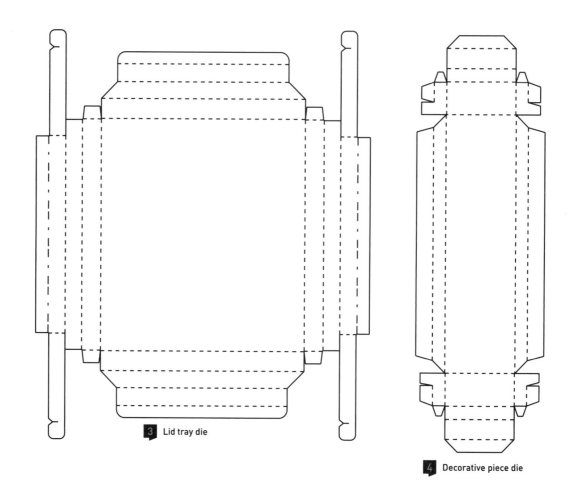

3 Lid tray die

4 Decorative piece die

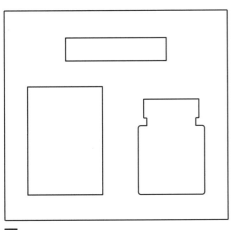

5 Base filling die, 20mm foam

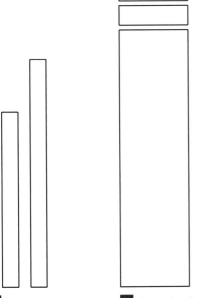

6 Tray filling die,
double-layer
corrugated
cardboard x2

7 Decorative piece
filling, double-layer
corrugated
cardboard x2

073

ADVENT
CALENDAR BOX

Use: Advent calendar for gifts during the countdown to Christmas. **Development:** Set of printed boxes in three different sizes that are grouped in a microchannel base with a leg for displaying the calendar. Graphics on both sides to provide a new design each day as the gifts are opened. **Material:** 300 g cardboard and 280 g micro-flute cardboard. → 2401

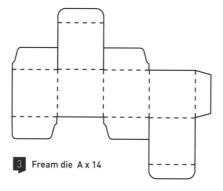

3 Fream die A x 14

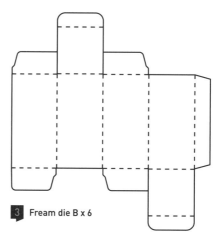

3 Fream die B x 6

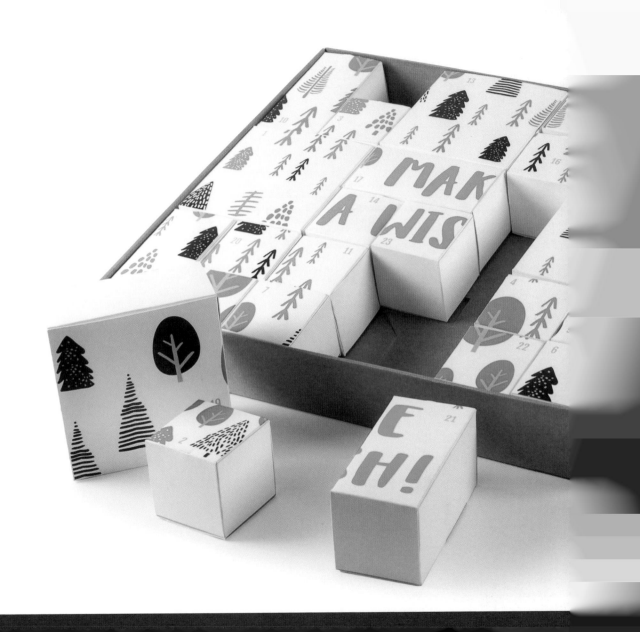

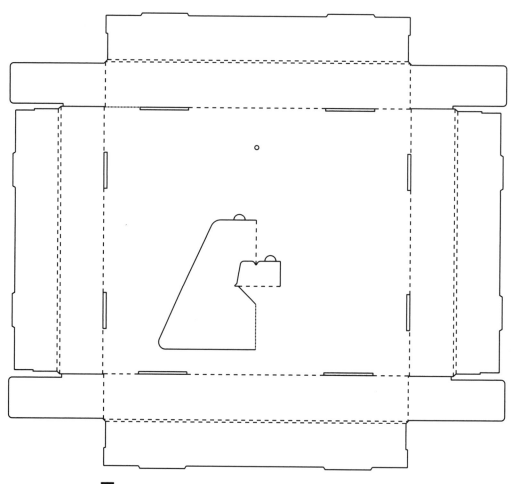

3 Base die

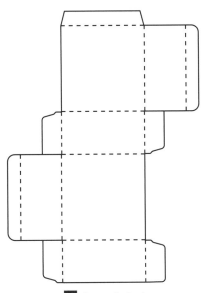

4 Fream die C x 4

074

INGOT-SHAPED BOX

Use: Box with sleeve, ideal for chocolates, candles or small objects. **Development:** Three different dies form this apparently simple box. The base, in the form of an ingot, includes a hollow space where an interior piece is placed to provide a better finish and where the objects are located. A piece in the form of a sleeve is used to close the pack. **Material:** 300 g paperboard. → 1800

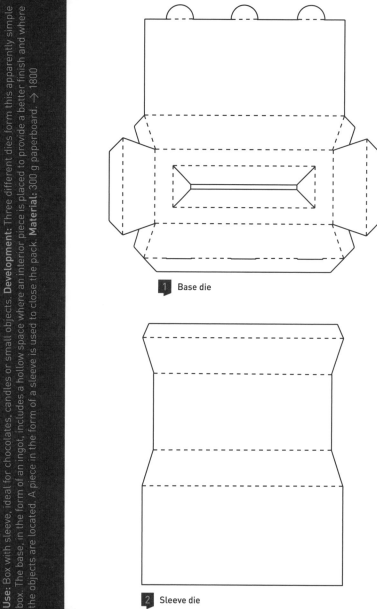

1 Base die

2 Sleeve die

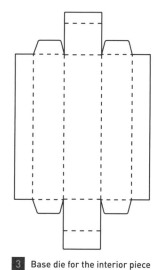

3 Base die for the interior piece

075

PRODUCT ON DISPLAY

Use: Two pieces joined together that leave one-fourth of the product visible. **Development:** Structure formed by two different-size rectangular pieces that form an "L" and are joined by a hinge to house the product, in this case, a Christmas bauble, suspended in the pack with only one-fourth of the ornament in sight. The die is adapted to the bauble's shape, protecting it while at the same time leaving it visible. The pack closes with magnets. **Materials:** 300 g paperboard, micro-flute cardboard, 18 mm magnet and strapping band. → 1640

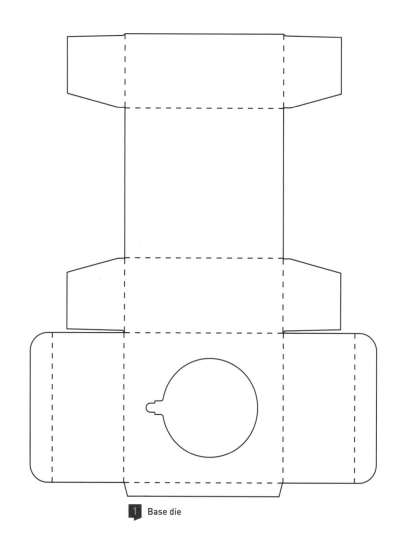

1 Base die

gasNatural UNION FENOSA

Las grandes tradiciones
se mantienen vivas y crecen
gracias a la innovación y a la
contribución de las personas
que las hacen posibles.

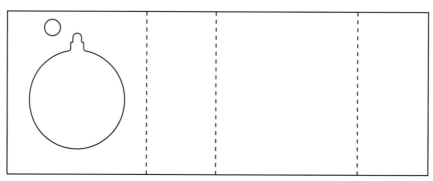

2 Base filling die, micro-flute cardboard

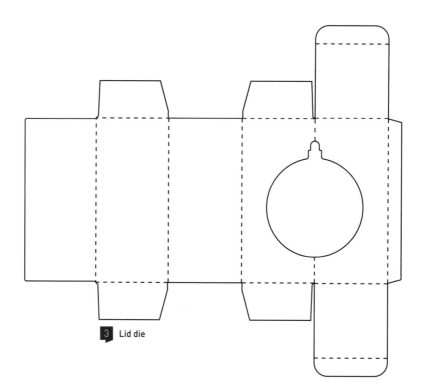

3 Lid die

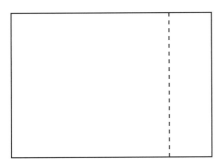

4 Hinge die

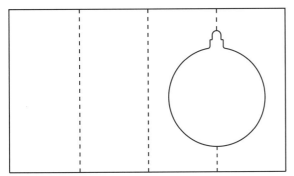

5 Filling die micro-flute cardboard

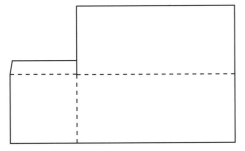

6 Interior filling die, micro-flute cardboard lid

076

SOMMELIER DISPLAY PACK

Use: Gift case for bottles and glass jar. **Development:** This packaging is formed by a perforated top and base to contain three fragile objects: two wine bottles and a glass jar, which are anchored to the base and remain in place thanks to an interior foam piece. The top is a tray that covers the product and base of the box completely. Two elastic bands straps ensure the pack closes properly. **Materials:** Double-layer corrugated cardboard, 300 g paperboard, 4 mm ø cord. → 1609

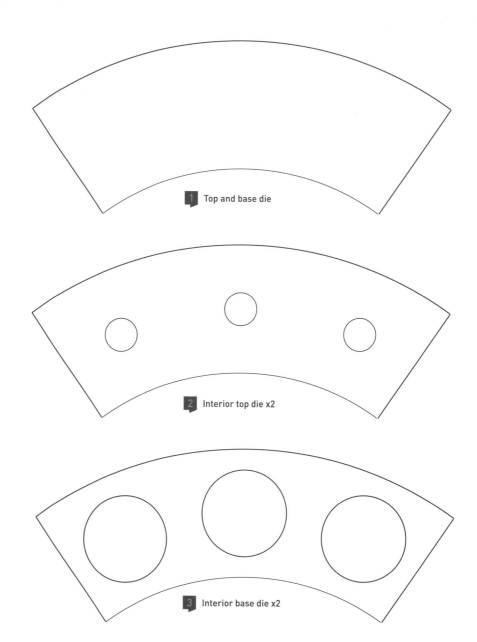

1 Top and base die

2 Interior top die x2

3 Interior base die x2

TRANSLUCENT CONNECTED PACK PACKAGE

Use: Box with visible brand and product. **Development:** Two-piece pack formed by a translucent plastic box that hints at the product inside and a cardboard piece in the shape of a "U". Both pieces snap into place, so the dies need to be adjusted perfectly. The plastic box is slightly raised, and therefore its contents shouldn't be too heavy. **Materials:** 280 g paperboard, double-layer corrugated cardboard and 0.5 mm polypropylene. → 0890

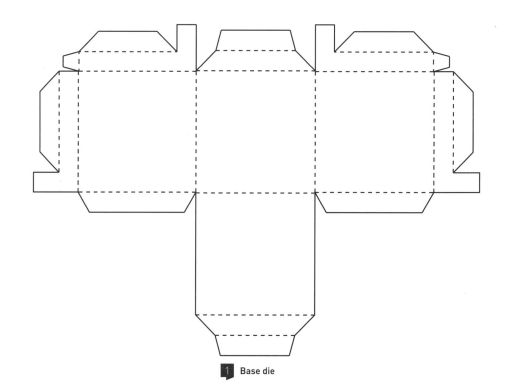

1 Base die

TRANSLUCENT
CONNECTED PACK PACKAGE

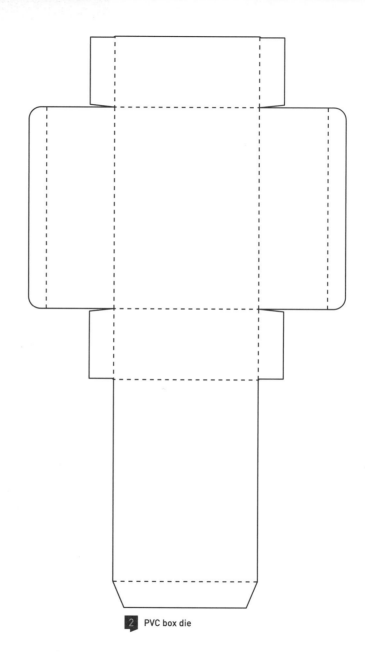

2 PVC box die

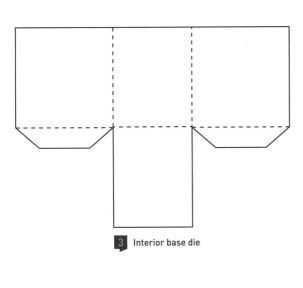

3 Interior base die

4 Base filling die, double-layer corrugated cardboard x2

5 Base filling die, double-layer corrugated cardboard x2

TRIPLE BUCKET PACK

Use: Box for presenting a product line. **Development:** Lined box comprising different levels that fit inside one another, thus creating detachable units and allowing each product to be stored according to the colour of the edging. **Materials:** 300 g paperboard, gray cardboard, cloth paper, 30 mm foam. → 2406

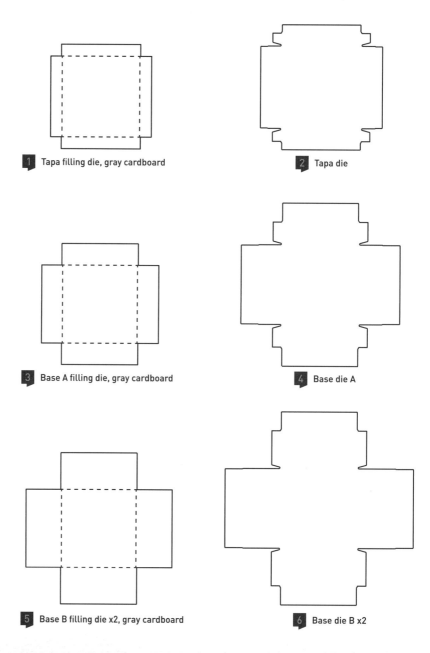

1 Tapa filling die, gray cardboard

2 Tapa die

3 Base A filling die, gray cardboard

4 Base die A

5 Base B filling die x2, gray cardboard

6 Base die B x2

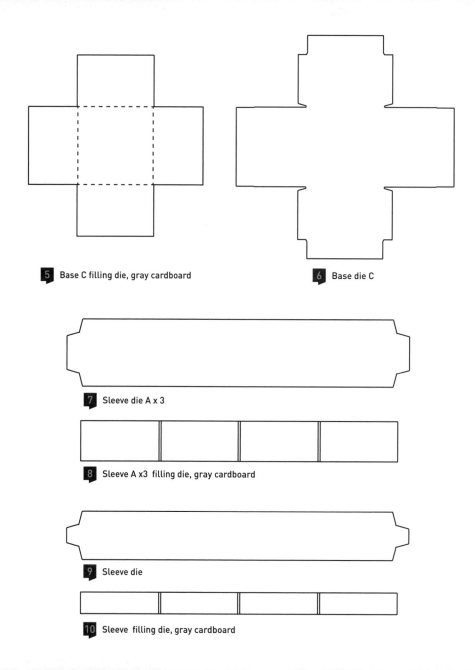

5 Base C filling die, gray cardboard

6 Base die C

7 Sleeve die A x 3

8 Sleeve A x3 filling die, gray cardboard

9 Sleeve die

10 Sleeve filling die, gray cardboard

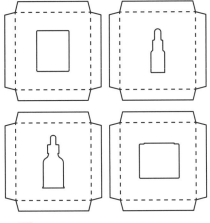

6 Interior die

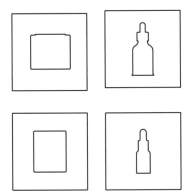

6 40 mm foam die

TRIPTYCH DISPLAY BOX

Use: Welcome pack for bottles that can be used as a display case. **Development:** This box is formed by three pieces that open in accordion fashion. The central piece, die-cut in the shape of a bottle, must be precisely adjusted to the bottle to prevent it from slipping and falling out when the pack is opened. The lateral parts are anchored to the box with a hinge and serve as a lid. A system of magnets is used for closing the pack. **Materials:** 300 g paperboard, Porexpan, micro-flute cardboard. → 1097

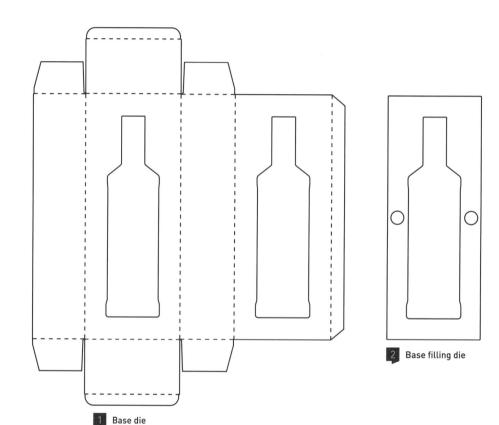

1 Base die

2 Base filling die

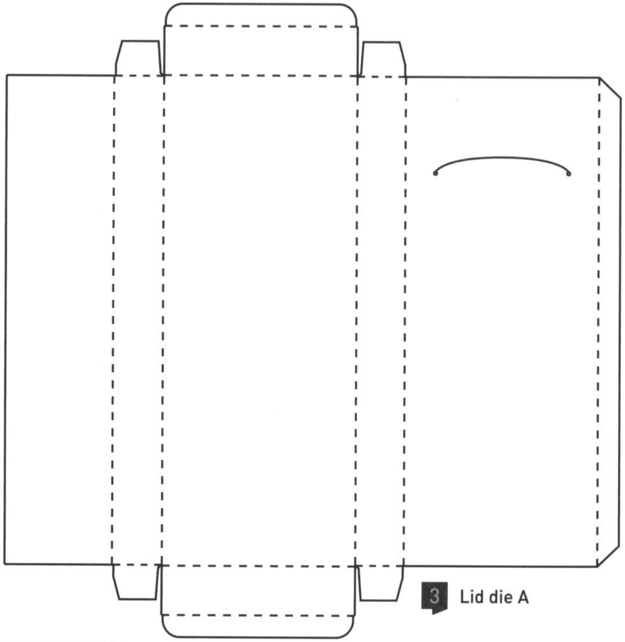

W

3 Lid die A

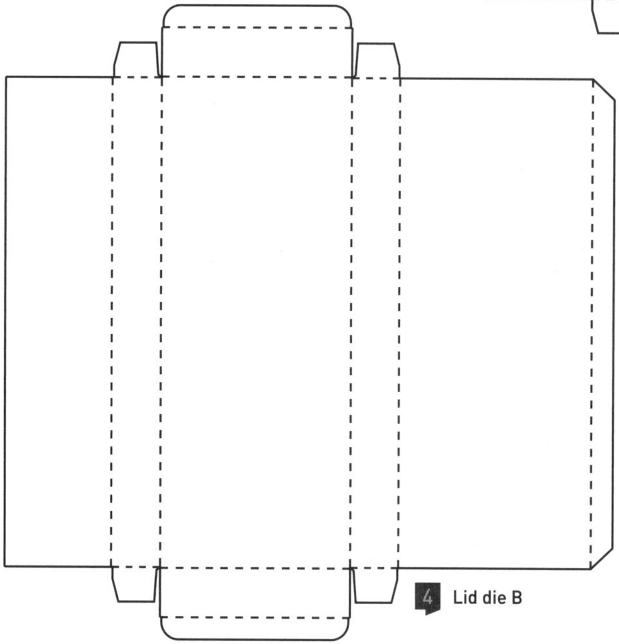

4 Lid die B

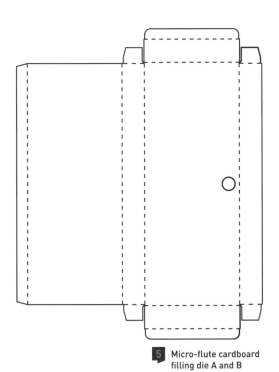

5 Micro-flute cardboard filling die A and B

080

ECOLOGICAL CHRISTMAS PACK

Use: Gift box that combines paperboard and double-layer corrugated cardboard. **Development:** The material of this gift box evokes the object it contains: a ginger root. A tray made from superimposed layers of double-layer corrugated cardboard where the root is placed forms the pack. The tray has a slot that fits the information brochure. The pack closes securely thanks to a U-shaped piece of cardboard that functions like a clamp. **Materials:** 300 g paperboard, double-layer corrugated cardboard. → 1641

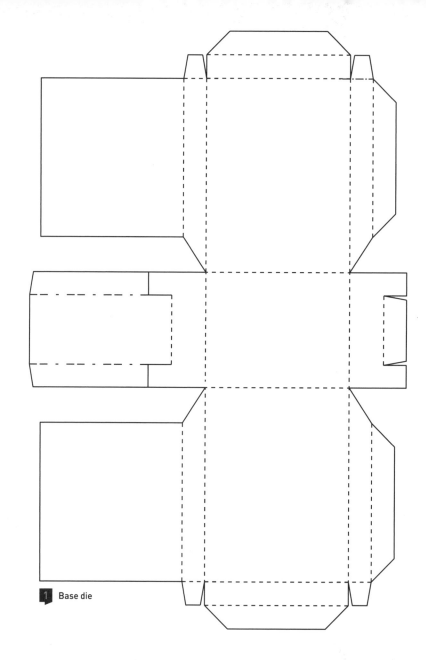

1 Base die

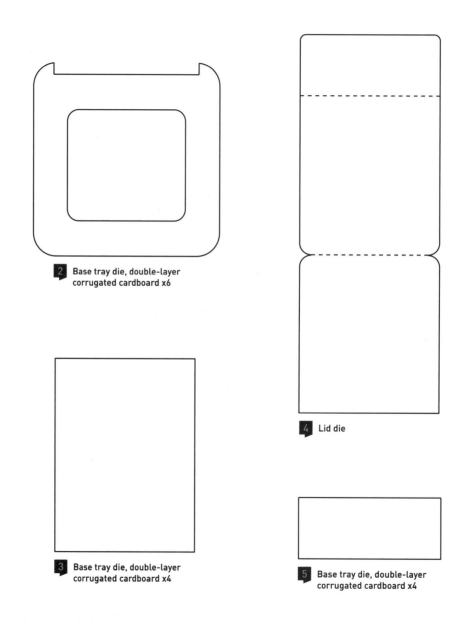

2 Base tray die, double-layer corrugated cardboard x6

4 Lid die

3 Base tray die, double-layer corrugated cardboard x4

5 Base tray die, double-layer corrugated cardboard x4

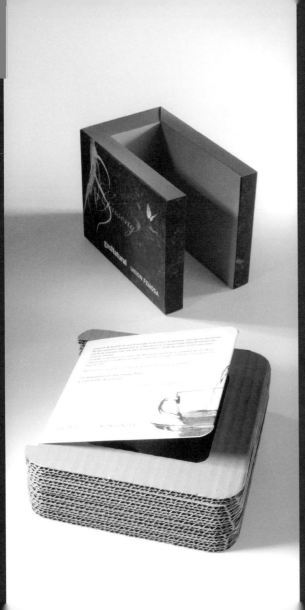

081

TWO
INTERLOCKING "L"s

Use: Pack formed by two symmetrical L-shaped boxes. **Development:** Two symmetrical "L"s joined in inverted fashion by a lateral hinge form this package. Each one has a flap at the end where the product is introduced. It is ideal for containing small, not too heavy objects. When closed, the box forms a perfect rectangle. **Material:** 300 g paperboard. → 1865

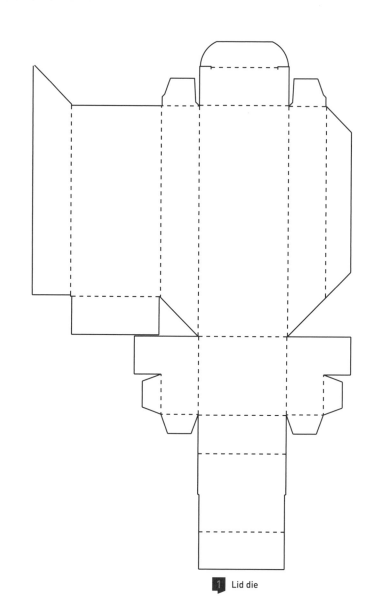

1 Lid die

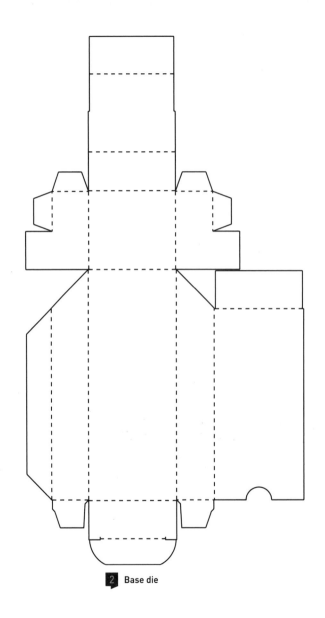

2 Base die

3 Interior base die

4 Interior base die

082

FOAM BOX
WITH HANDLE

Use: Display box with a handle. **Development:** This box consists of three dies: one, which is larger, has a handle, making the box easy to carry. The pieces are cut in the shape of the product the pack contains, adapting perfectly to its outline. As a finishing touch, a see-through sleeve envelops the box. **Material:** 10 and 20 mm foam and 0.5 mm polypropylene. → 1056

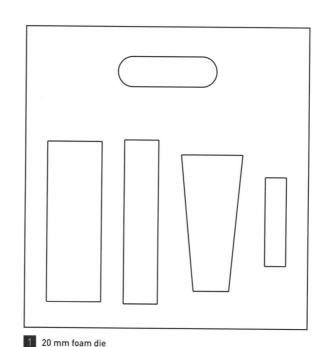

1 20 mm foam die

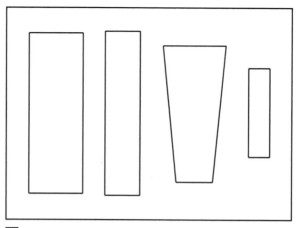

2 10 mm foam die

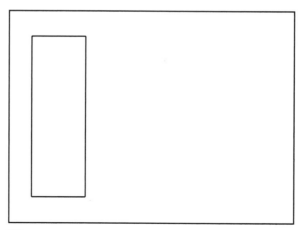

3 10 mm foam die

4 Sleeve die

083

MINIMALIST DISPLAY CASE

Use: This box combines foam and methacrylate to create an elegant product display case. **Development:** The packaging incorporates different materials in black tones, combining methacrylate elements with EVA foam and black paperboard. The silkscreen texts on the pack grant it a touch of elegance while the notably minimalist aesthetic highlights the product. **Materials:** 300 g paperboard, 3 mm methacrylate, 35 mm foam and 18 mm ø magnet. → 2353

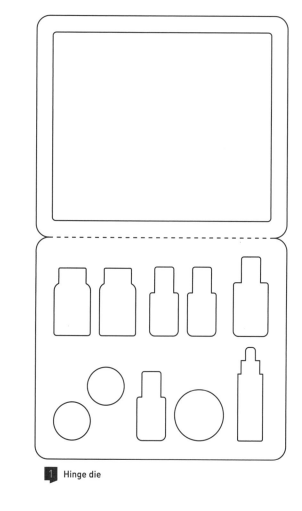

1 Hinge die

2 Foam lid die

3 Methacrylate lid die

4 Foam base die

5 Methacrylate base die

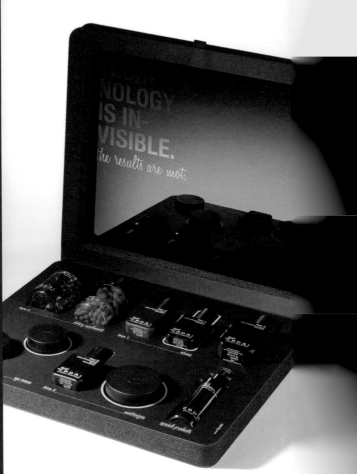

084

DOUBLE SYMMETRY BOX

Use: Gift kit with a surprise effect opening. Self-assembly. **Development:** This simple looking box is made up of several dies that allow for a surprise opening. The base piece includes sides that help keep the lids slightly raised when opening the pack. The product is fixed to the base by a perforated interior piece. **Materials:** 300 g paperboard, double-layer corrugated cardboard. → 2334

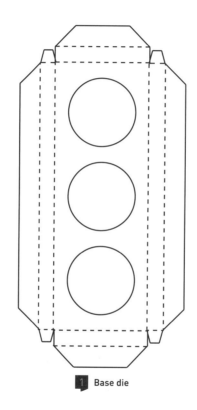

1 Base die

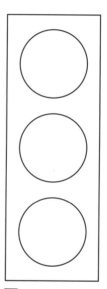

2 Base filling die, double-layer corrugated cardboard

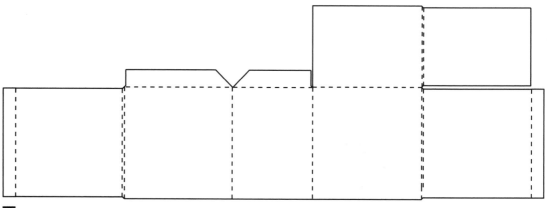

 Lid die A

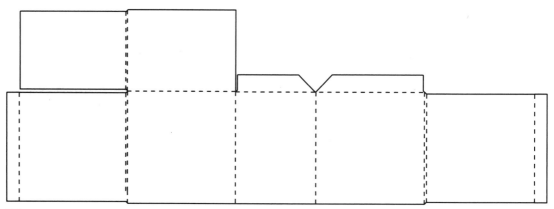 **Lid die B**

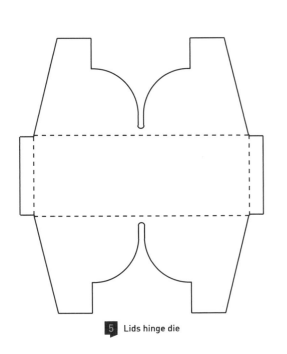

5   Lids hinge die

085

DOUBLE CUBE PACK

Use: Gift case formed by two pieces that frame the product. Self-assembly. **Development:** This box highlights the product inside, in this case, two cosmetic bottles. It consists of two pieces that serve as a frame thanks to the die-cut in the shape of each bottle. It closes by means of a sleeve with the company name or logo printed on it. **Material:** 300 g paperboard. →2176

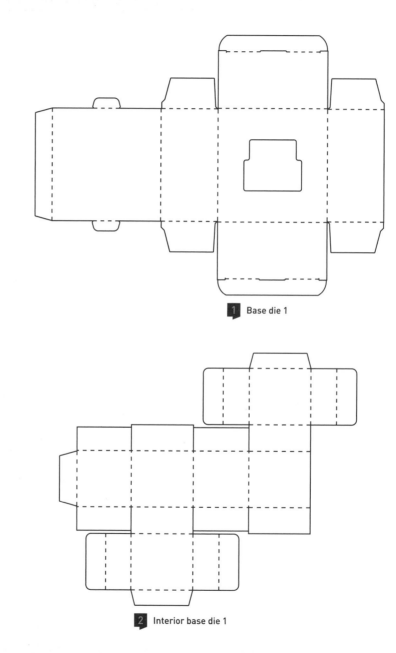

1 Base die 1

2 Interior base die 1

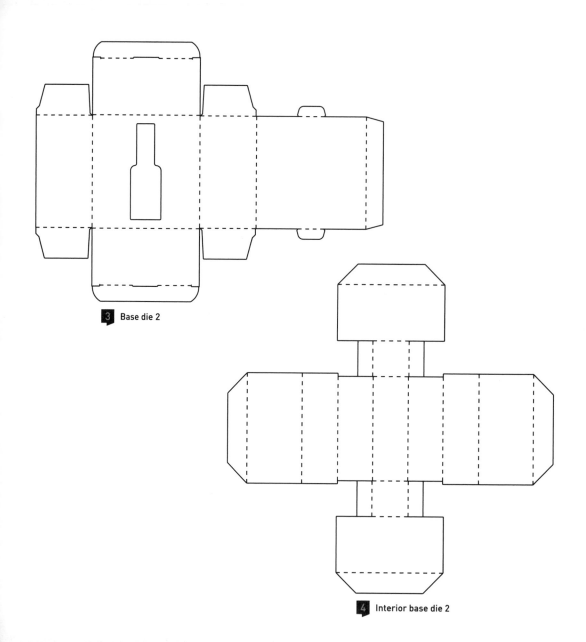

3 Base die 2

4 Interior base die 2

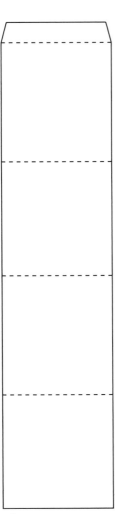

 Sleeve die

086

GIFT KIT
SAMPLE CASE

Use: Welcome pack for a book and product samples. **Development:** This welcome pack includes a series of holes for placement of the product samples, in addition to a space for a book or information brochure. The base is die-cut with spaces specifically designed for the objects it is intended to contain, adjusting completely to their shape. It closes thanks to a simple printed lid. **Materials:** 150 and 300 g paperboard, 20 mm and 28 mm foam and 2 mm gray cardboard. → 2329

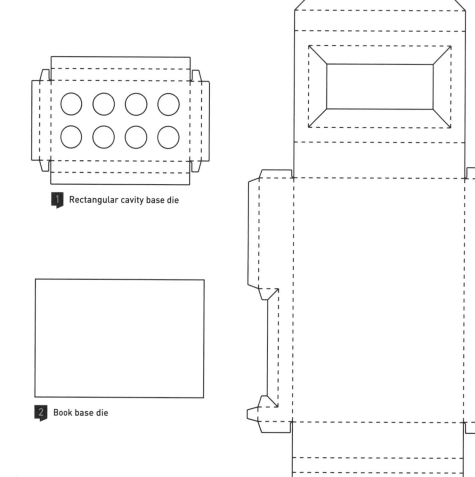

1 Rectangular cavity base die

2 Book base die

3 Base die

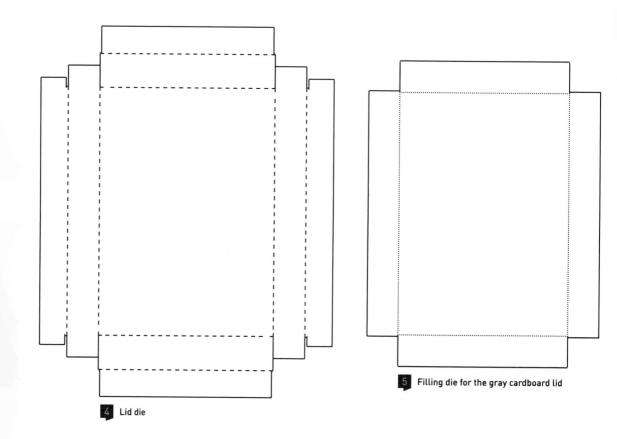

4 Lid die

5 Filling die for the gray cardboard lid

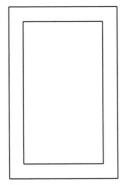

6 28 mm foam filling die

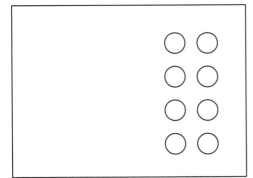

7 20 mm foam filling die

087

PERFORATED SQUARE PACK

Use: Pack for containing spherical objects. **Development:** Formed by three foam pieces joined by a cardboard hinge. The three pieces have a central hole with different diameters. This allows the object to fit into the central piece in such a way that the other two pieces, of a lesser diameter, leave the object in sight while at the same time holding it perfectly in place. An adhesive that joins the pieces closes the pack securely. **Materials:** 20 mm and 40 mm foam, 280 g paperboard. → 0738

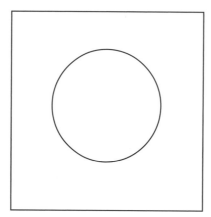

1 Interior base die

2 Closing die

3 Base die x2

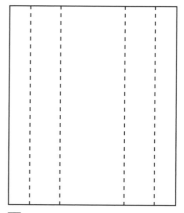

4 Hinge die

088

FRAMED
GIFT BOX

Use: This pack is actually an ideal frame for enhancing a product. **Development:** While the structure of the pack is simple, it serves as an effective support for the wine glass it contains. A single die that frames and protects the product forms the outer piece. An interior piece secures the glass in place by its top part. Interior supports provide added strength to the pack. **Materials:** 200 g paperboard, micro-flute cardboard. → 2337

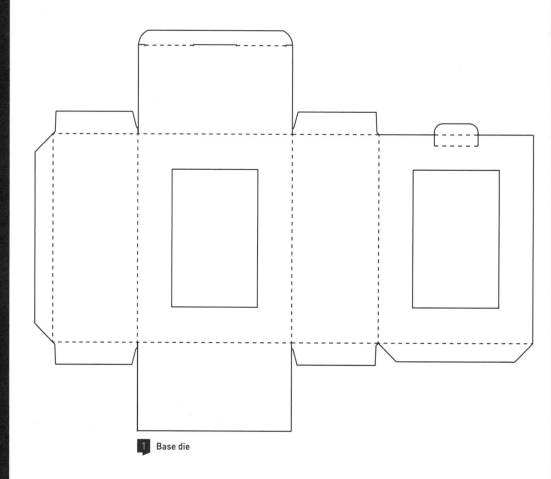

1 Base die

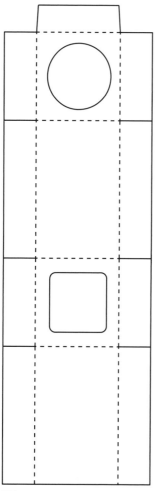

2   Interior base die

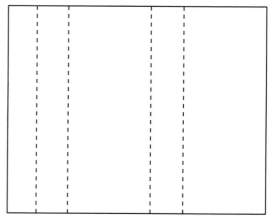

3   Base filling die, micro-flute cardboard x2

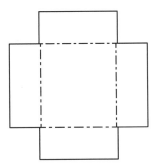

4 Base filling die, micro-flute cardboard

5 5 mm foam filling die

/252/ /256/ /300/ /268/ /272/ /380/ /280/

/376/ /396/ /304/ /336/ /316/ /352/ /296/

/332/ /368/ /320/ /344/ /360/ /348/ /308/

/364/ /384/ /392/ /372/ /276/ /324/ /312/

/284/ /264/ /292/ /340/ /260/

/356/ /328/

/288/ /388/

/LEVEL03/

CAR
WELCOME PACK

Use: Display case. **Development:** This diptych box is formed by two pieces joined at the spine by a hinge. The first piece is a frame reinforced with micro-flute cardboard that contains a miniature car. The box is perforated, leaving the car visible on both sides of the frame. The other piece is a small case with a CD and brochure, with a thumb-notch for easy removal. The pack is closed with a tray in the form of a lid that also includes a thumb-notch opening. **Materials:** 300 g paperboard and 2 mm gray cardboard and a foam CD button. → 1745

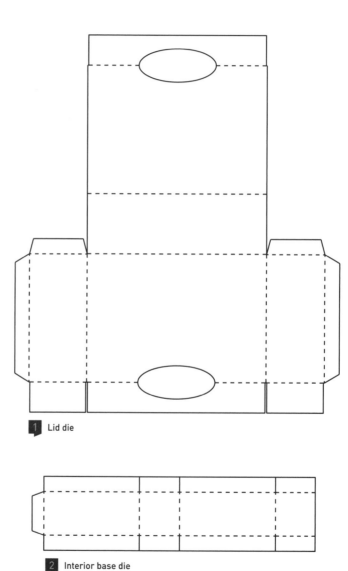

1 Lid die

2 Interior base die

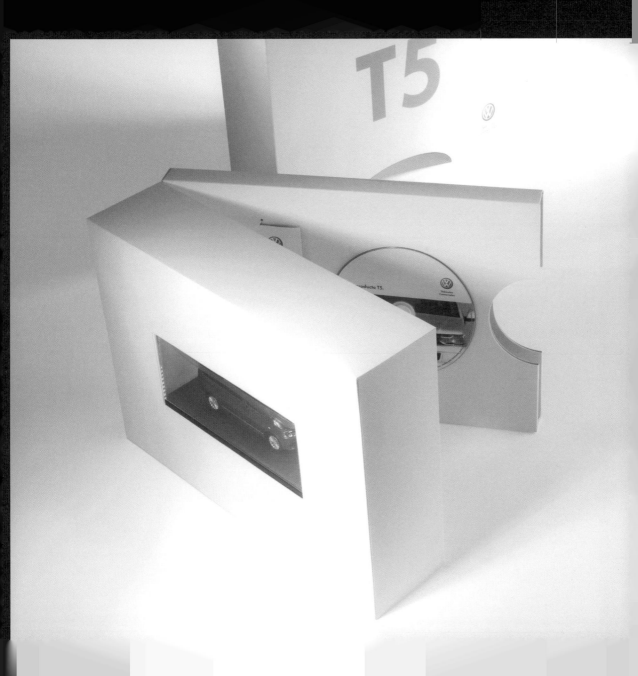

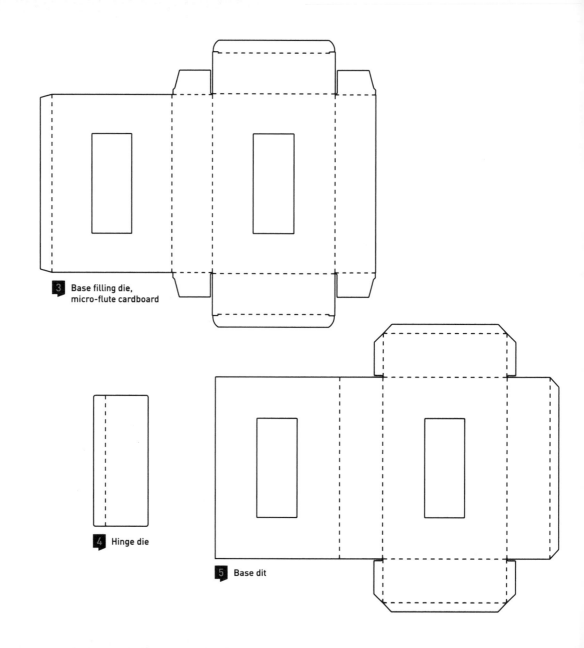

3 Base filling die, micro-flute cardboard

4 Hinge die

5 Base dit

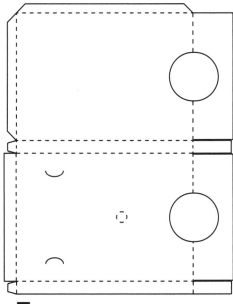

6 Pocket die

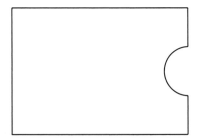

7 Gray cardboard pocket filling die

090

ROUNDED-EDGE CHEST

Use: To highlight and protect a glass bottle. **Development:** Interior made with micro-flute cardboard pieces arranged one on top of the other to protect and support the bottle. Other vertically placed pieces reinforce this rounded shape, creating a structure similar to the one used in old wooden boats. Magnets are hidden inside to ensure proper closing. **Materials:** 300 g paperboard, 18 mm magnet, double-layer corrugated cardboard, 18 mm satin ribbon. → 2159

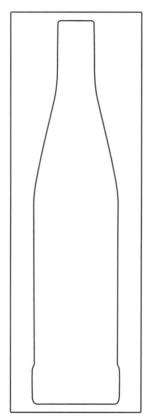

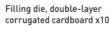 Filling die, double-layer corrugated cardboard x10

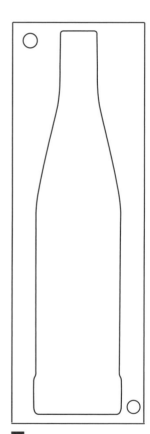

2 Base die, double-layer corrugated cardboard x2

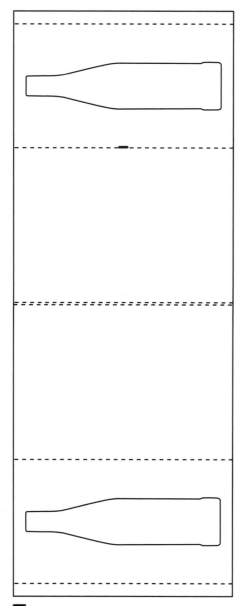

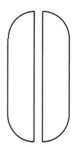

4 Filling die, double-layer corrugated cardboard x2

5 Filling die, double-layer corrugated cardboard x6

3 Base die

091

MOBILE
DISPLAY FRAME

Use: Display case for 4 product bottles. **Development:** This multi-use box in the form of a frame is designed to hold 4 bottles and dramatically highlight the product. The frame is divided into two pieces connected by magnets that are separated to remove the bottles. The bottles fit into holes in the top and bottom of the frame that hold the product in place and, thanks to an interior foam layer, are perfectly adjusted to the product's shape. **Materials:** 300 g paperboard, double-layer corrugated cardboard, 18 mm ø magnet and 10 mm foam. → 1986

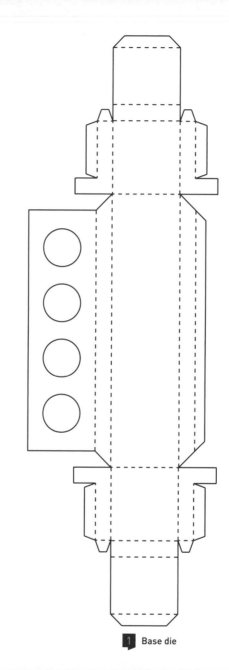

1 Base die

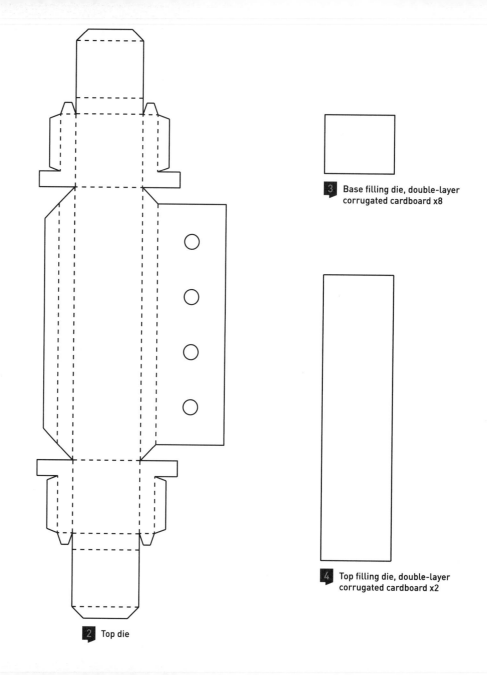

3 Base filling die, double-layer corrugated cardboard x8

4 Top filling die, double-layer corrugated cardboard x2

2 Top die

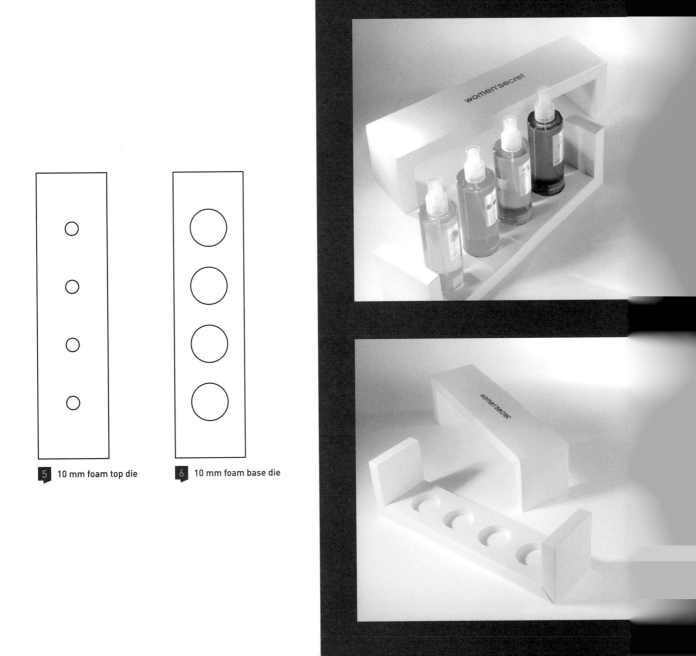

5   10 mm foam top die

6   10 mm foam base die

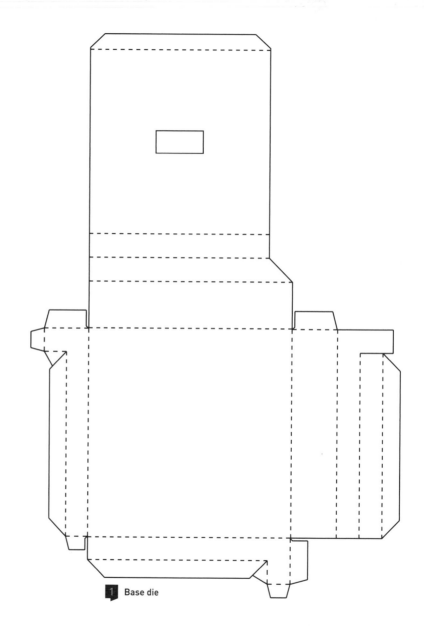

1 Base die

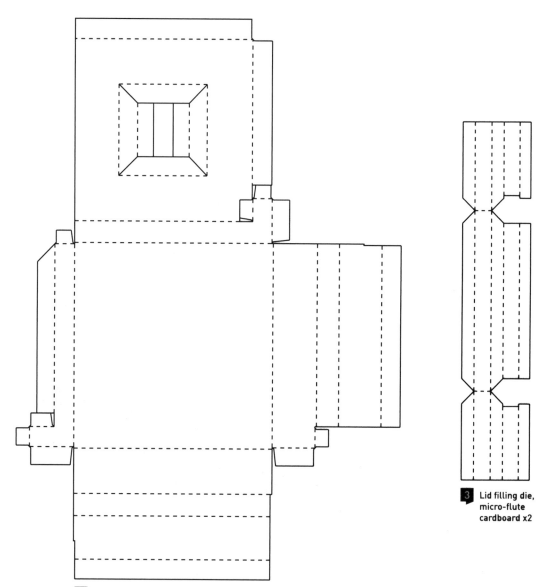

2 Base filling die, micro-flute cardboard

3 Lid filling die, micro-flute cardboard x2

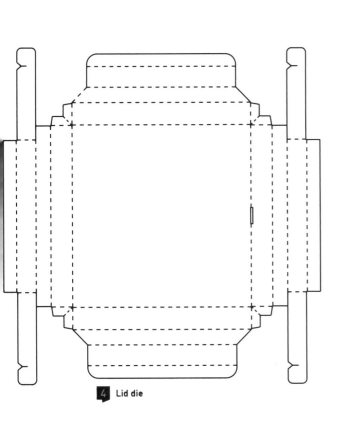

4 Lid die

093

MULTIPLE ACCORDION BOX

Use: Multi-product presentation with six trays. **Development:** Six trays joined by five hinges form the pack. In three of them, foam interiors hold three perfume bottles in place. This interior can be adapted to any object keeping in mind its thickness. The product boxes, which can be removed, fit in the other three trays. When the pack is closed, the bottles don't touch each other. The box closes using a conventional lid. **Materials:** 300 g paperboard, 40 mm foam. → 1984

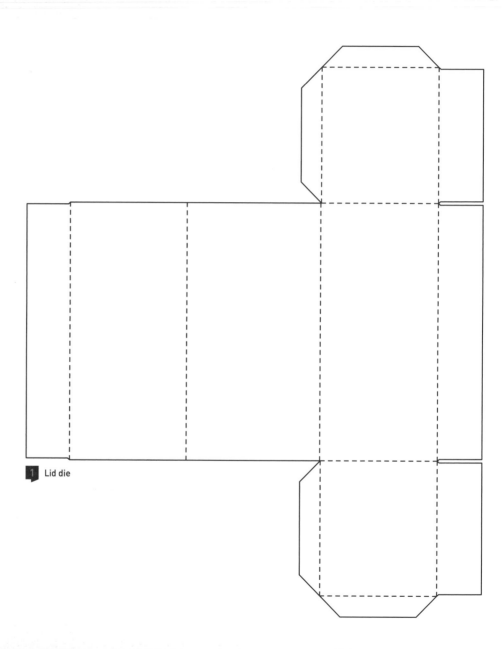

1 Lid die

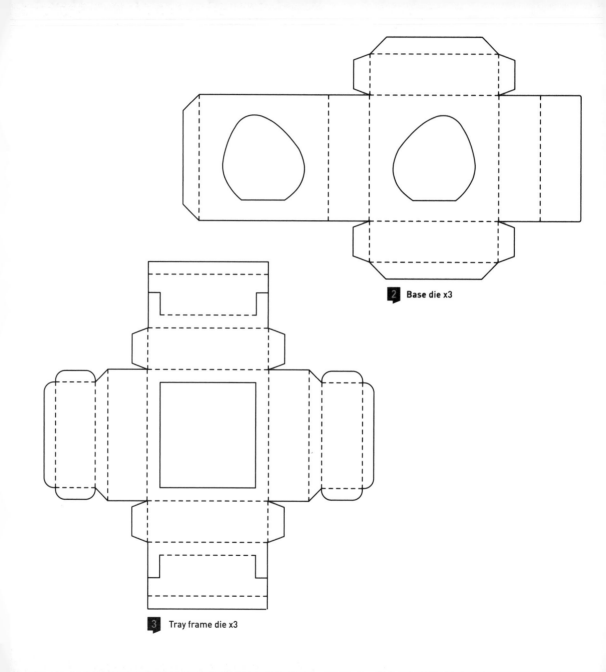

2 **Base die x3**

3 Tray frame die x3

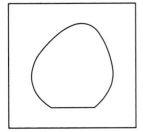

4 40 mm foam filling die x3

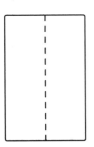

5 Hinge die x5

094

DIAGONAL SURPRISE PACK

Use: Display window pack. **Development:** The sides of the base and lid of this pack have been cut on an angle to form an inclined plane when opening it. An inner filling stabilizes the base when the bottle is removed. The top turns 180° when the pack is opened, coming to rest diagonal to the base to form an attractive display. The height of the object needs to be kept in mind to ensure that the lid closes properly. It is recommended to add an adhesive fastener. **Materials:** 300 g paperboard, micro-flute cardboard. → 1643

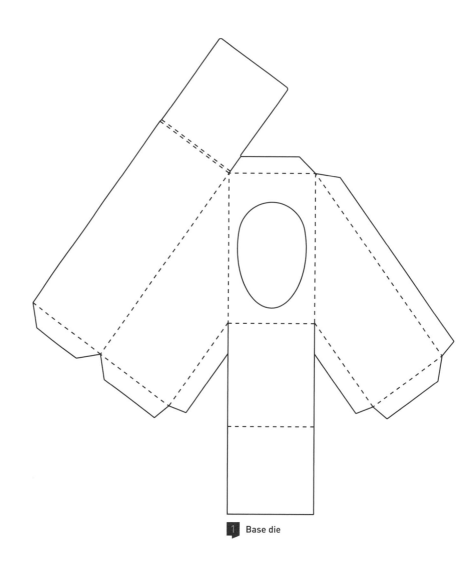

1 Base die

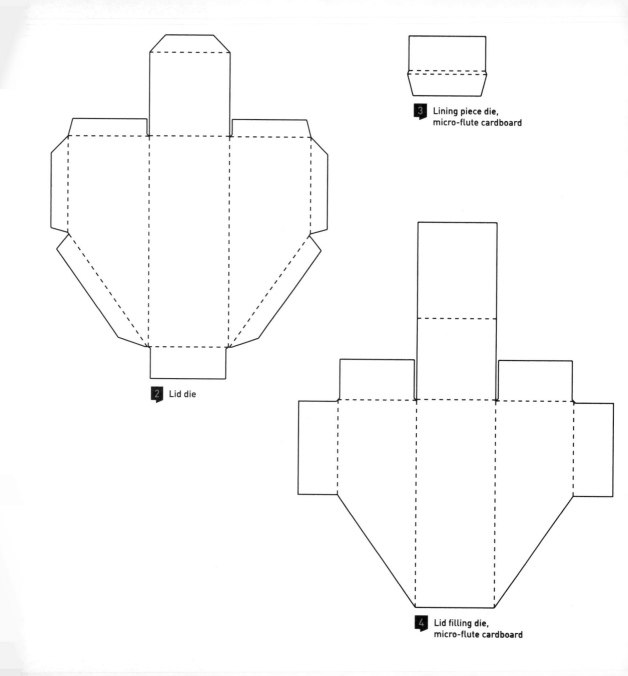

3 Lining piece die, micro-flute cardboard

2 Lid die

4 Lid filling die, micro-flute cardboard

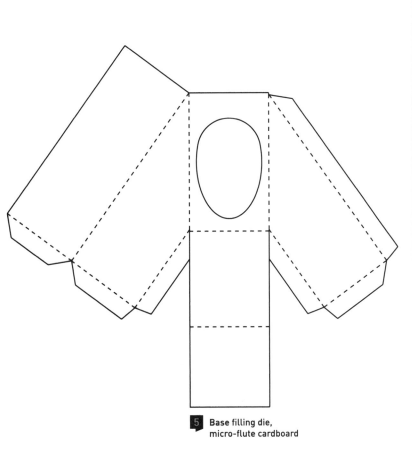

5 **Base** filling die,
micro-flute cardboard

095

FEMININE FRAGRANCE WELCOME PACK

Use: A graceful system for displaying a women's perfume. **Development:** Packaging with an interior created from a circular foam block both for the base and the lid, to which we have glued printed paperboard. For closing the box, the base has double-thick foam adjusted perfectly to the empty space in the lid. The outer bow has no practical function; it is a decorative element that adheres to the aesthetic of the brand. **Materials:** 300 g paperboard, 45 mm and 20 mm foam, 30 mm wide ribbon. → 2181

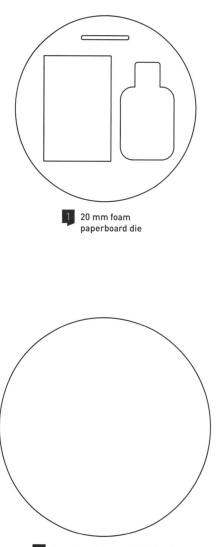

1   20 mm foam paperboard die

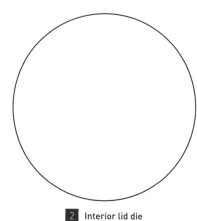

2   Interior lid die

3   Paperboard base and lid die x2

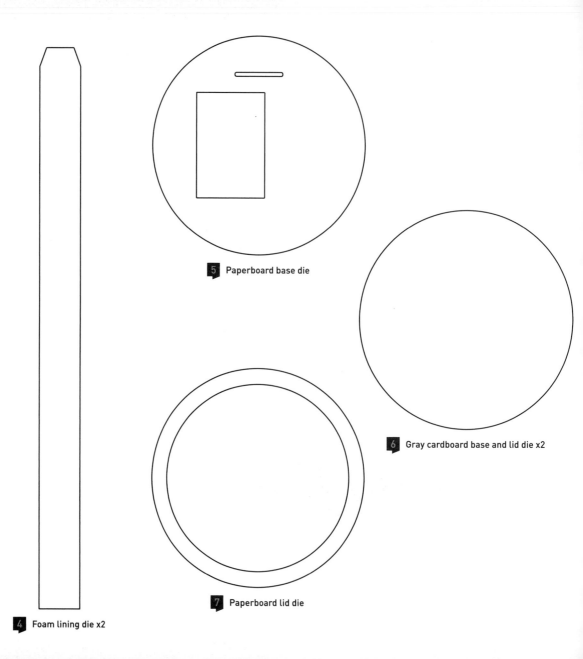

5 Paperboard base die

6 Gray cardboard base and lid die x2

7 Paperboard lid die

4 Foam lining die x2

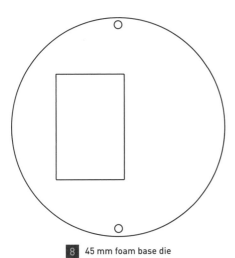

8 45 mm foam base die

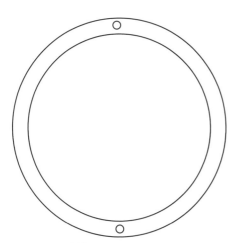

9 45 mm foam lid die

096

NEW FRAGRANCE DISPLAY

Use: Welcome pack. **Development:** This packaging is made up of 4 pieces: a base, two lids and a methacrylate piece that protects and highlights the bottle. Cardboard hinges that support the lids are added to the base. The die of the small lid was reduced to the same thickness as the methacrylate to ensure the pack closes perfectly. **Materials:** 300 g paperboard, 4 mm methacrylate, 30 mm foam, 3 mm ø and 18 mm ø magnets, 18 mm wide ribbon, 3 mm ø elastic band. → 2366

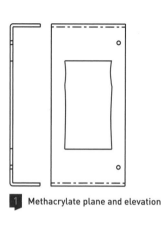

1 Methacrylate plane and elevation

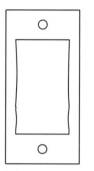

2 30 mm foam filling die

3 Hinge die

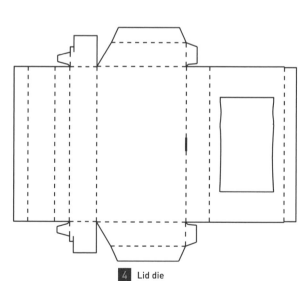

4 Lid die

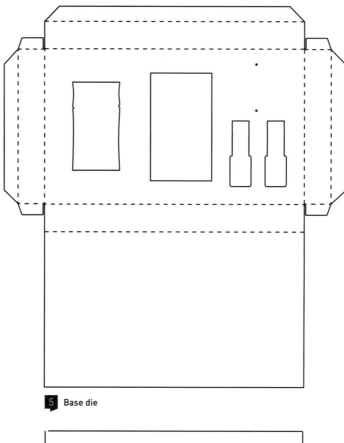

5 Base die

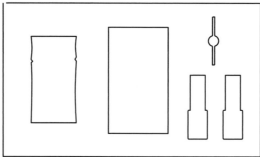

6 30 mm foam filling die

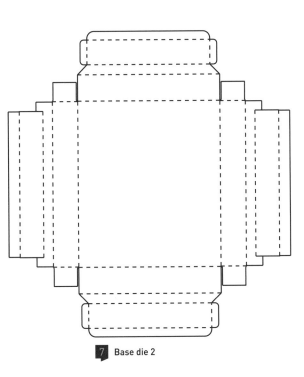

7 Base die 2

DISPLAY TRAY
WELCOME PACK

Use: Box designed as a display case for small product line samples. **Development:** This pack with simple, minimalist lines is intended to highlight the box's content: a series of product samples. The base is a vertical tray whose borders frame and highlight the product. It includes a lid whose inner face contains an information card. A strip of fabric was added to facilitate opening. **Materials:** 300 g paperboard, double-layer corrugated cardboard, 18 mm ø magnet, 15 mm foam. → 2386

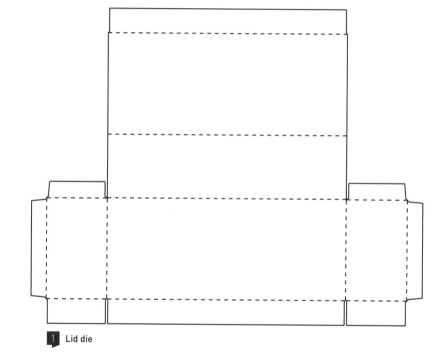

1 Lid die

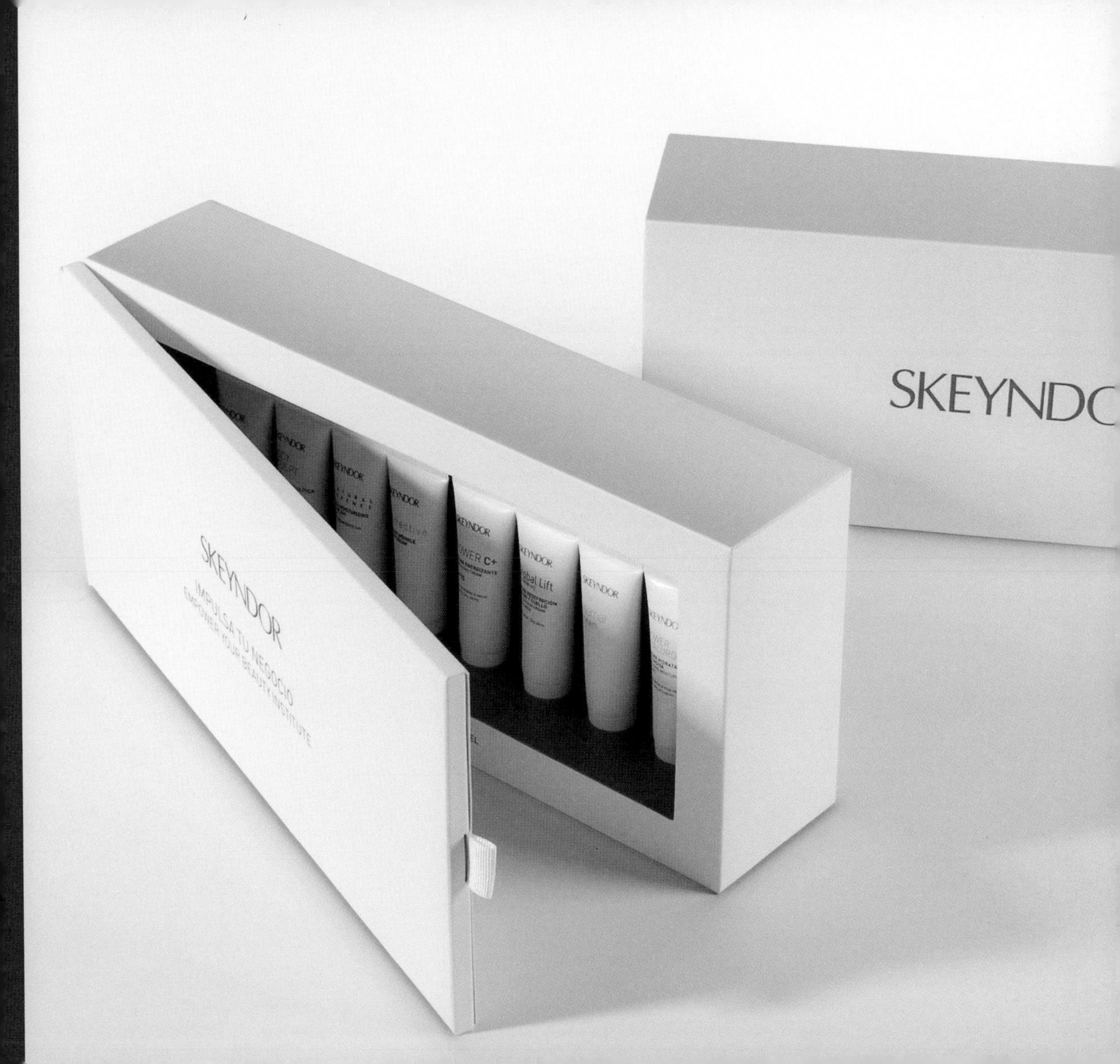
SKEYNDOR
IMPULSA TU NEGOCIO
EMPOWER YOUR BEAUTY INSTITUTE

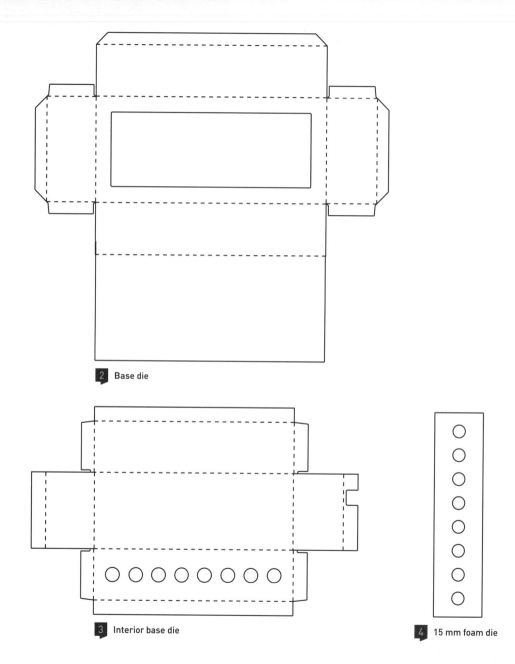

2 Base die

3 Interior base die

4 15 mm foam die

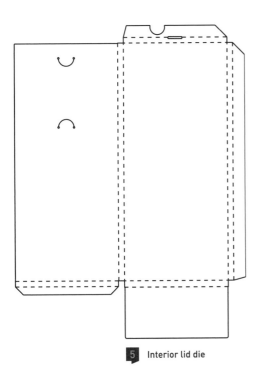

5 Interior lid die

6 Interior lid filling die, double-layer corrugated cardboard

098

CHRISTMAS TREE BOX

Use: Gift box for a Christmas lamp. **Development:** The interior of this pack is die-cut in the form of a Christmas tree, as it was designed to contain a lamp in this shape. The pack consists of two rectangular pieces: one is a rectangular tray where the object is inserted while the other has an interior die in the shape of the lamp. When closing the pack, this allows half of the lamp to remain visible. **Materials:** 0.8 mm polypropylene, 300 g paperboard, double-layer corrugated cardboard, bulb socket and bulb. → 1985

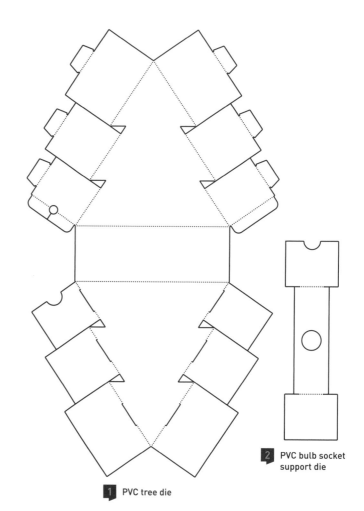

1   PVC tree die

2   PVC bulb socket support die

Navidad
2010

En Gas Natural Fenosa sabemos que la
energía no se crea ni se destruye sino que
se transforma. Y por eso dedicamos
nuestros esfuerzos a transformar esa
energía en bienestar y confort, con la
máxima eficiencia energética y respeto
por el medio ambiente.

Le invitamos a pasar una Navidad más
eficiente, iluminando estas fiestas con
energía de bajo consumo y materiales
ecológicos, con un único objetivo:
construir juntos un futuro más sostenible.

¡Felices Fiestas!

gasNatural
fenosa

Por un 2011
lleno de luz
y de calor

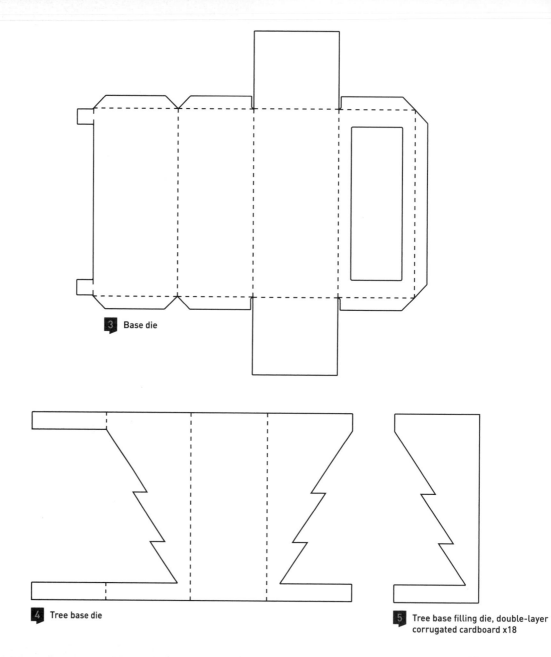

3 Base die

4 Tree base die

5 Tree base filling die, double-layer corrugated cardboard x18

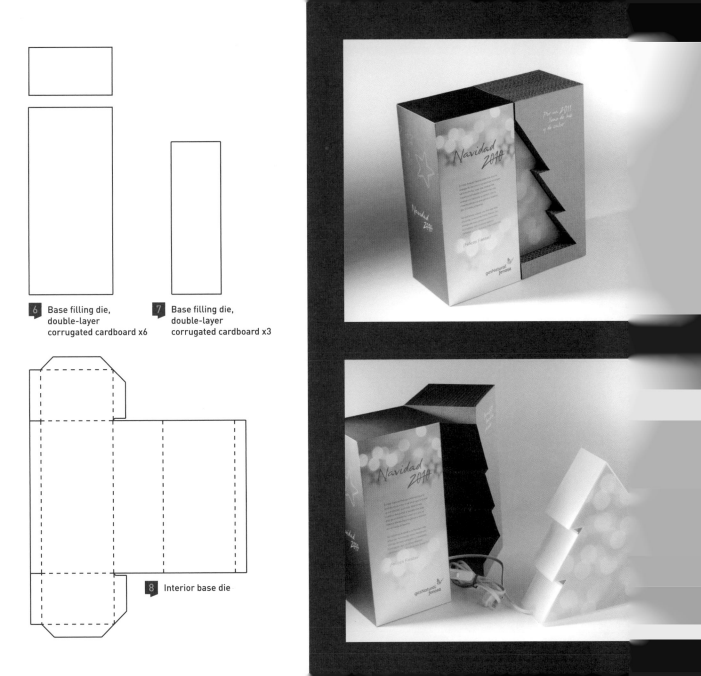

6 Base filling die,
double-layer
corrugated cardboard x6

7 Base filling die,
double-layer
corrugated cardboard x3

8 Interior base die

099

COFFEE MAKER DISPLAY PACK

Use: Welcome pack. **Development:** The base of the pack contains three product samples displayed on a piece of methacrylate and covered with a lid without revealing the rest of the contents. A secondary compartment contains the coffee maker, also covered with a lid. A box-bag with handles makes the pack easy to carry. **Materials:** 300 g paperboard, 40 mm foam, micro-flute cardboard, 10 mm methacrylate, 18 mm ø magnet, 15 mm fabric ribbon. → 1908

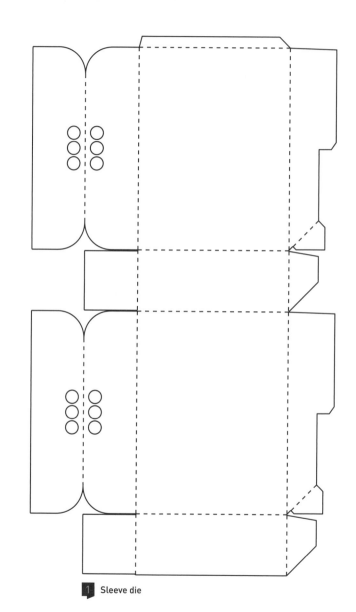

1 Sleeve die

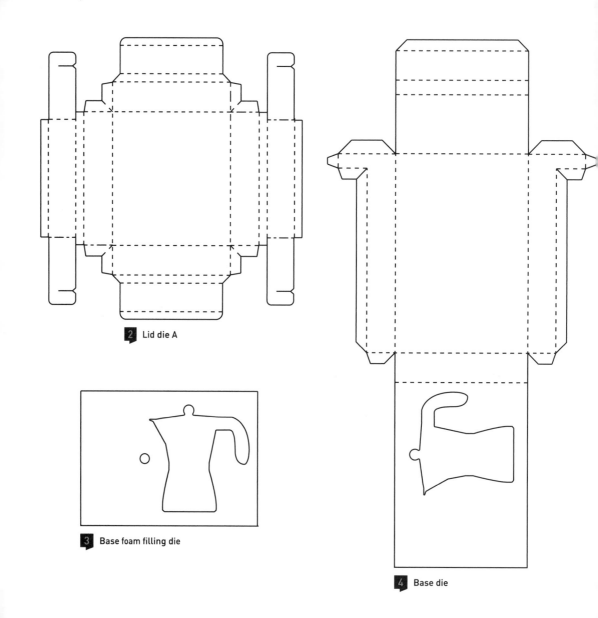

2 Lid die A

3 Base foam filling die

4 Base die

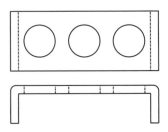

5 Methacrylate plane and elevation

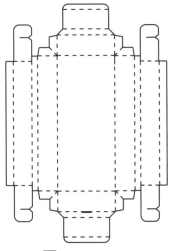

6 Lid die B

7 Hinge die

METHACRYLATE BOX

Use: Elegant box that combines cardboard and methacrylate. **Development:** The methacrylate base includes several circular holes to fit the product units, which partially stick out. The lid is filled with foam and covers the top of the product. Magnets connect the lid and base to close the pack. **Materials:** 300 g paperboard, 32 mm methacrylate, 2 mm gray cardboard, 5 mm and 10 mm Ø magnets, strapping band, 10 mm foam and 0.3 mm PET plastic. → 2153

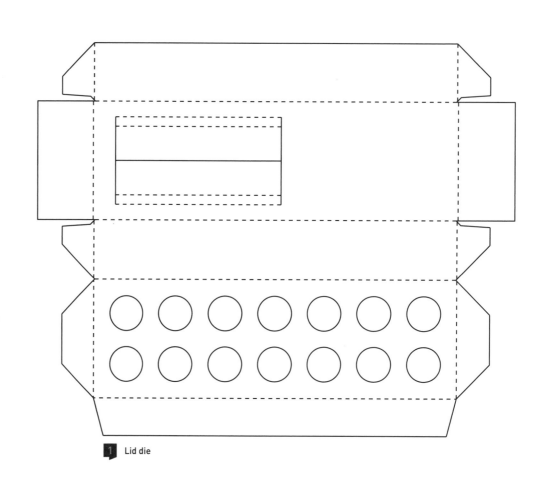

1 Lid die

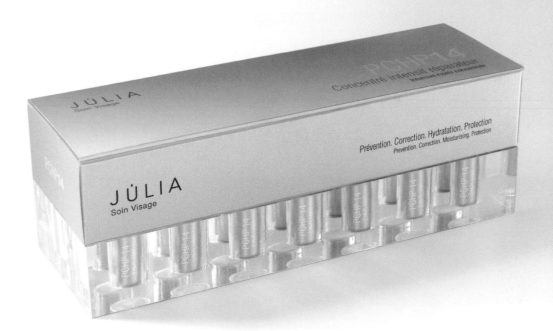

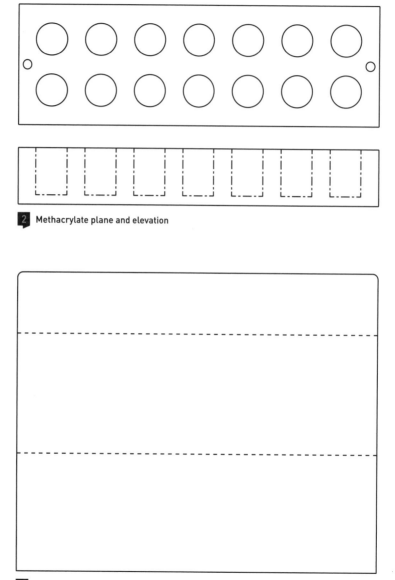

2 Methacrylate plane and elevation

3 Lid die

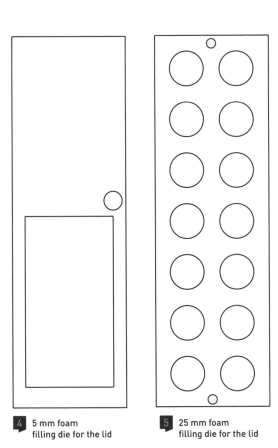

4 5 mm foam
filling die for the lid

5 25 mm foam
filling die for the lid

101

PROMOTIONAL CASE PACK

Use: Jewellery box for watches. **Development:** Double-function box. Removing the interior piece, the box can be reused as a jewel case for watches. It is important to keep in mind that the height and thickness of the interior pieces and the width of the bottle must have the same dimensions as the interior of the two closed trays for the bottle to be properly secured. A cardboard sleeve is used to close the pack. **Materials:** 300 g paperboard, micro-flute cardboard, 30 and 40 mm foam, 10 mm ø magnet and strapping band. → 1015

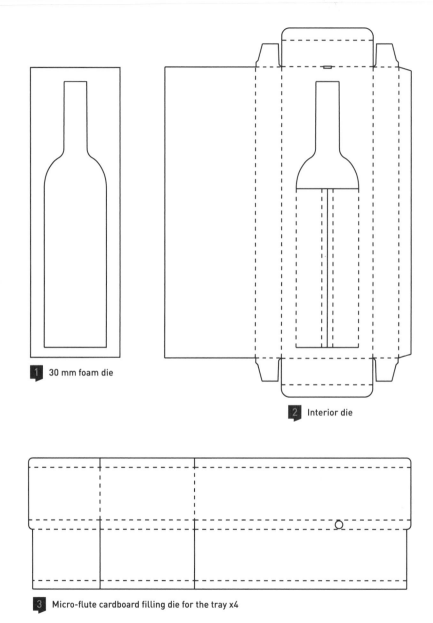

1. 30 mm foam die

2. Interior die

3. Micro-flute cardboard filling die for the tray x4

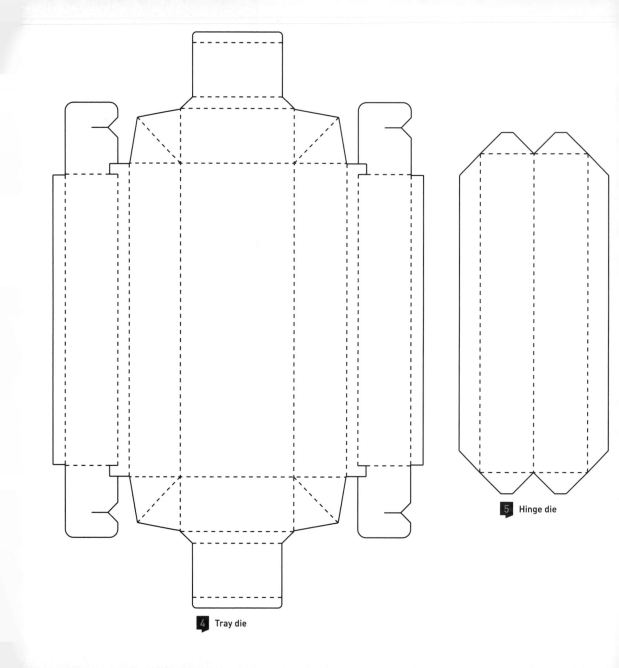

4 Tray die

5 Hinge die

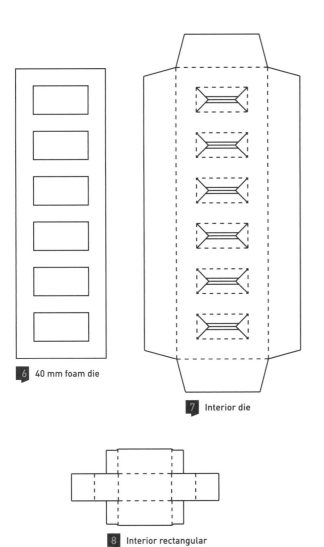

6 40 mm foam die

7 Interior die

8 Interior rectangular cavity die x6

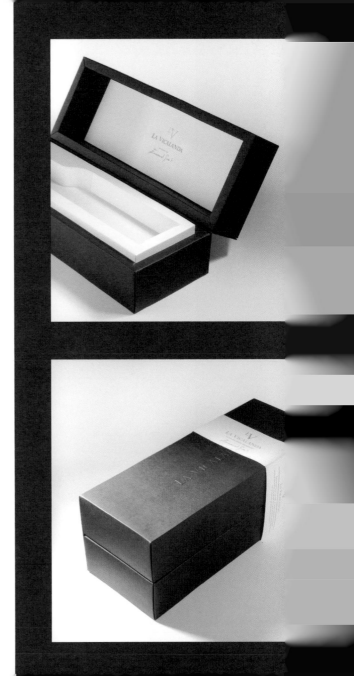

102

DISPLAY
TRAY INSERT

Use: Product display box. **Development:** This box is formed by two U-shaped pieces that dovetail on top of each other. The thickness of both pieces needs to be kept in mind so that, applying light pressure, the box closes securely. The pack has two uses: on the one hand it presents the product while on the other it can serve as a display arrangement if the lid is used as a base. **Materials:** 300 g paperboard, micro-flute cardboard. → 1098

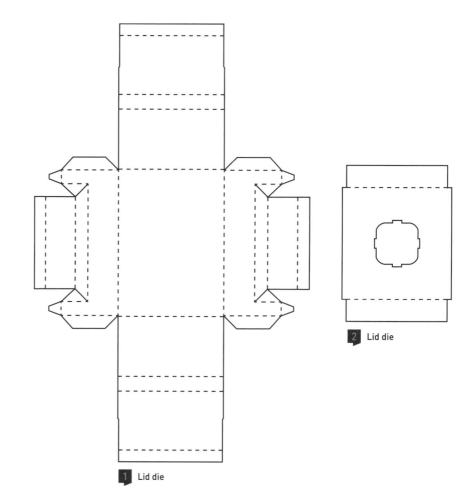

1 Lid die

2 Lid die

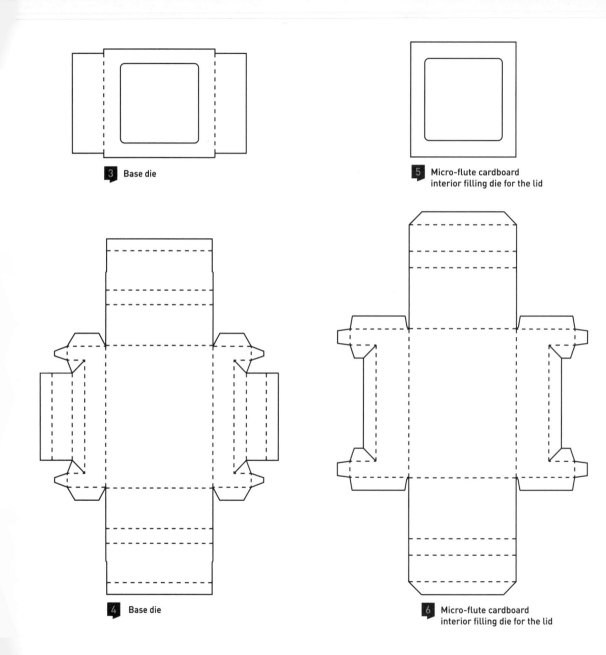

3 Base die

5 Micro-flute cardboard interior filling die for the lid

4 Base die

6 Micro-flute cardboard interior filling die for the lid

7 Micro-flute cardboard
interior filling die for the base

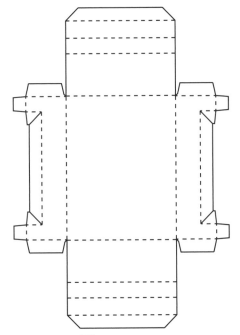

8 Micro-flute cardboard
interior filling die for the base

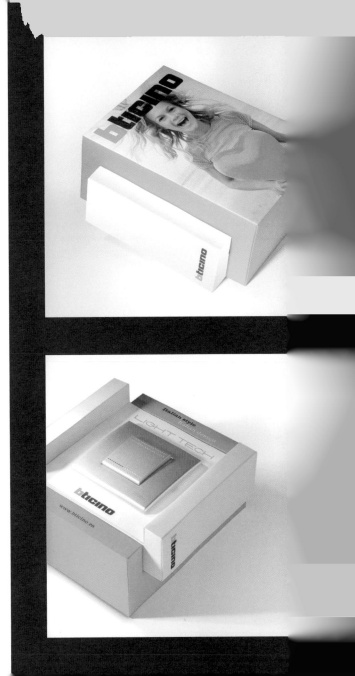

PLUSH
WELCOME PACK

Use: Welcome pack. **Development:** The first thing about this pack that grabs our attention is the layer of white plush. We use this material in allusion to the name of the Skoda Yeti car model. The pack has two trays: one with gadgets that allude to the car's characteristics and another that contains information material about the car: a CD and a brochure. **Materials:** 300 g paperboard, micro-flute cardboard, plush fabric, small methacrylate boxes, 18 mm magnet and strapping band. → 1215

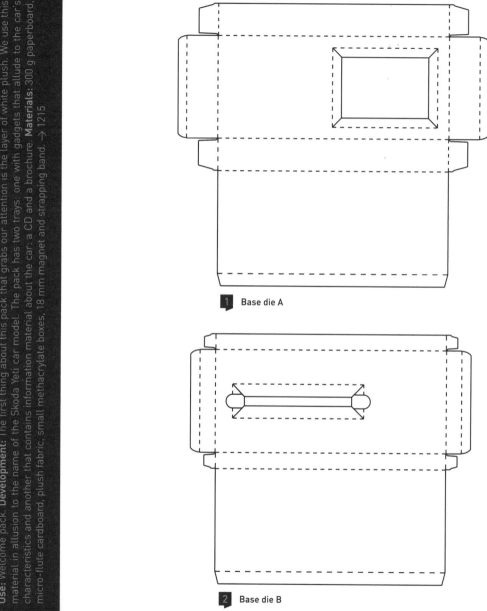

1 Base die A

2 Base die B

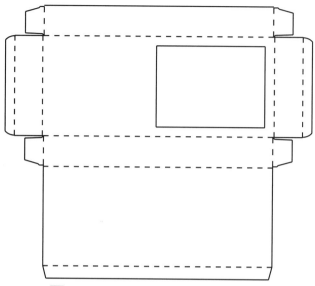

3 Base A filling die, micro-flute cardboard

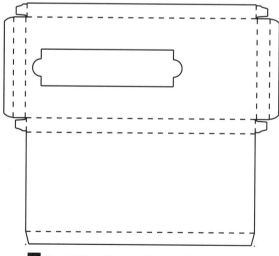

4 Base B filling die, micro-flute cardboard

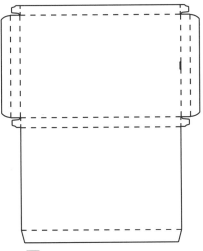

5 Lid die

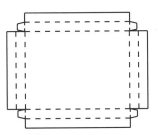

6 Rectangular cavity die for base A

7 Hinge die

104

WOMEN'S PERFUME CHEST

Use: Welcome pack in the shape of a chest. **Development:** The base is made from a conventional box with a foam block inside to protect the fragile contents. The lid is constructed out of pieces of rounded foam, lined with paperboard, which is used as a hinge and joins the lid and base. Inside are hidden several magnets to ensure proper closing. **Materials:** 300 g paperboard, 25 mm and 50 mm foam, double-layer corrugated cardboard, tulle fabric, 4 18 mm ø magnets, 3 mm ø elastic band and flocking. → 2301

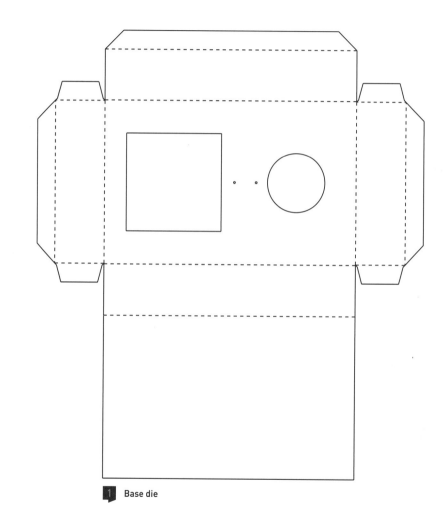

1 Base die

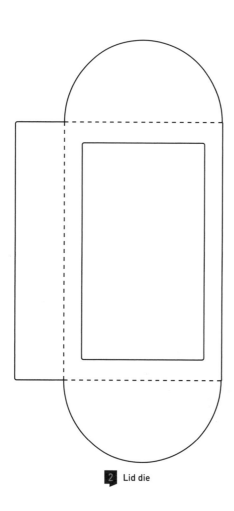

2 Lid die

3 Flocking die

4 25 mm foam filling die for the lid x 10

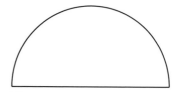

5 Filling die for gray cardboard lid and double-layer corrugated cardboard x4

6   Lid die

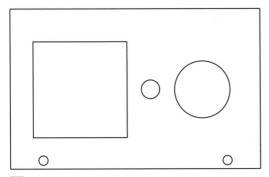

7   50 mm foam filling die for the base

105

ENCASED
BOTTLE PACK

Use: Box whose design and material protects fragile objects inside. **Development:** The box consists of two L-shaped parts that open to reveal the product and close with the help of magnets. It combines two styles: a smooth exterior cover and an interior made from double-layer corrugated cardboard to protect the bottle. The pack is a nice combination of rustic and elegant qualities. **Materials:** 300 g paperboard, double-layer corrugated cardboard, 10 mm ø magnet and strapping band. → 1227

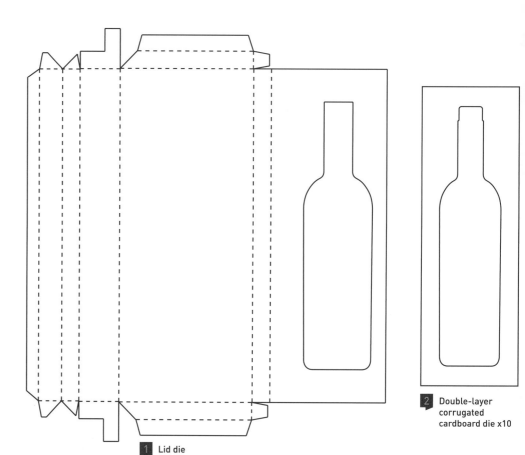

1 Lid die

2 Double-layer corrugated cardboard die x10

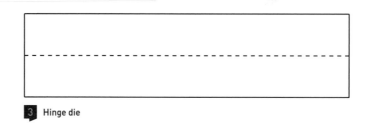

3 Hinge die

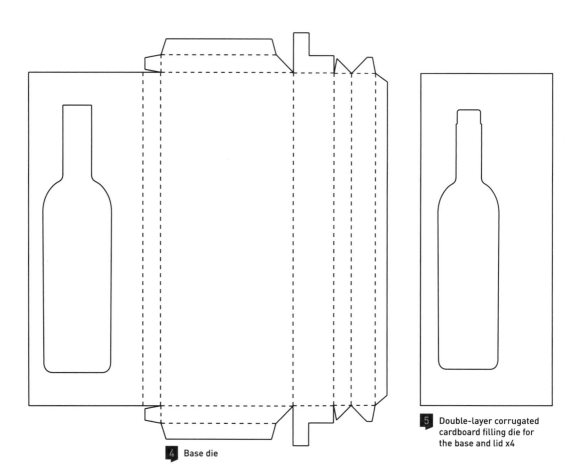

4 Base die

5 Double-layer corrugated cardboard filling die for the base and lid x4

6 Double-layer corrugated cardboard filling die for the base and lid x2

7 Double-layer corrugated cardboard filling die for the base and lid x6

106

W

Use: Transmit a service via packaging. **Development:** Packaging constructed out of two pieces each with three sides and that when joined, form an irregular cube, creating an interesting effect. The piece that we use as the base has a hole adjusted to the exact size of the gadget to prevent the latter from moving. The pack is closed with powerful magnets hidden inside the cardboard. **Materials:** 300 g paperboard, micro-flute cardboard, 18mm ø magnet, 15 mm foam. → 1327

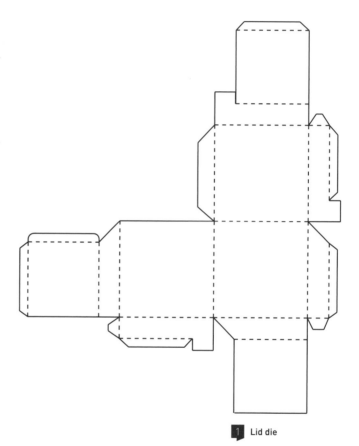

1 Lid die

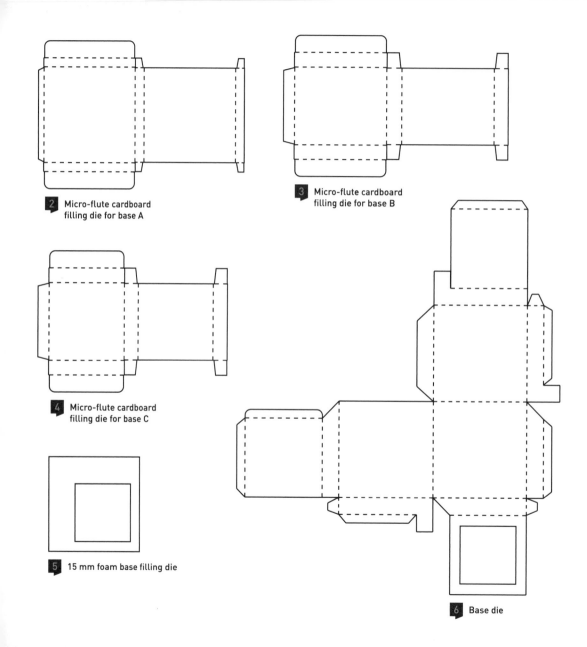

2 Micro-flute cardboard filling die for base A

3 Micro-flute cardboard filling die for base B

4 Micro-flute cardboard filling die for base C

5 15 mm foam base filling die

6 Base die

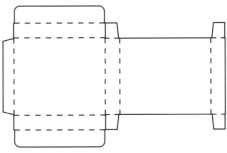

7 Micro-flute cardboard filling die for lid A

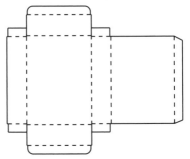

8 Micro-flute cardboard filling die for lid B

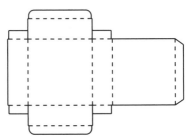

9 Micro-flute cardboard filling die for lid C

107

Use: Press kit. **Development:** This box is constructed around EVA foam, which allows for the creation of sinuous curves at the pack's corners. The cardboard lid and base are glued inside. The box closes through the use of magnets embedded in the foam. Also, the U-shaped base of the box enables the lid to fit perfectly in it. **Materials:** 300 g paperboard, 2 mm EVA foam, 10 mm ø magnets, strapping band and double-layer corrugated cardboard. → 0819

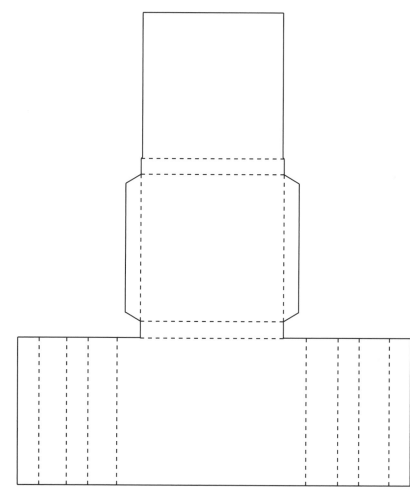

 Micro-flute cardboard filling die for the lid

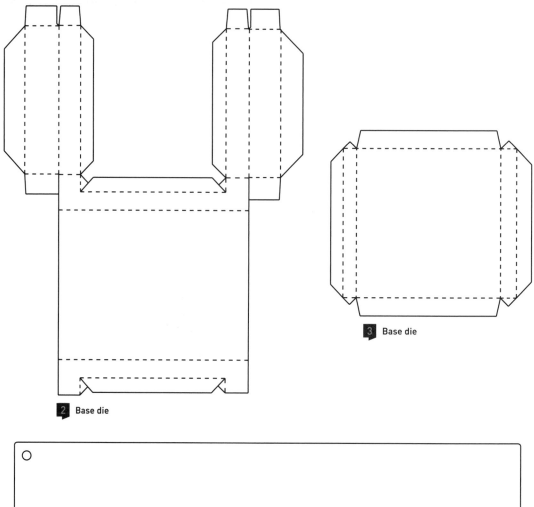

2 Base die

3 Base die

4 2 mm EVA foam die for the lid

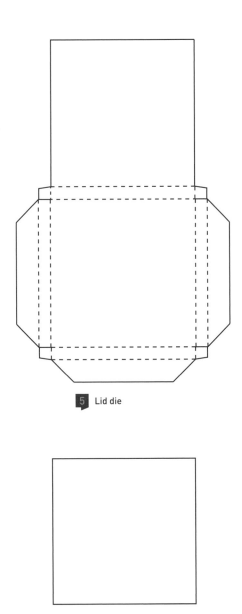

5 Lid die

6 Double-layer corrugated cardboard filling die for the lid x2

108

SLOPED BOX

Use: Irregular box to contain a product. **Development:** Complex box that plays with irregular lines. It incorporates double-layer corrugated cardboard interior pieces that provide strength to the pack while a U-shaped piece supports the product. The box closes with magnets and includes a cloth strap to facilitate opening. One end of the irregular opening lid must touch the ground to serve as a support when the pack is open. **Materials:** 300 g paperboard, micro-flute cardboard, 18 mm magnet and strapping band. → 0666

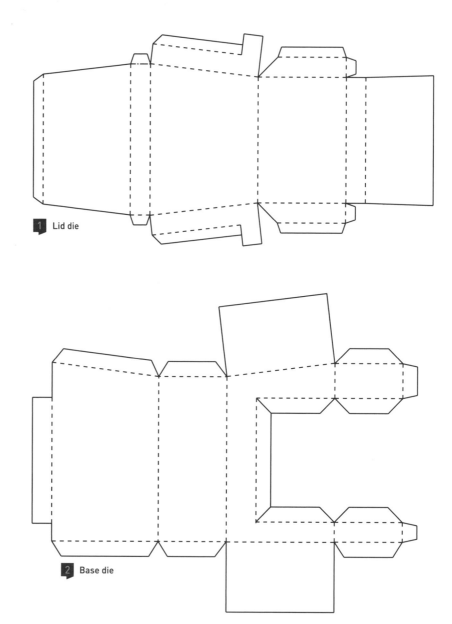

1 Lid die

2 Base die

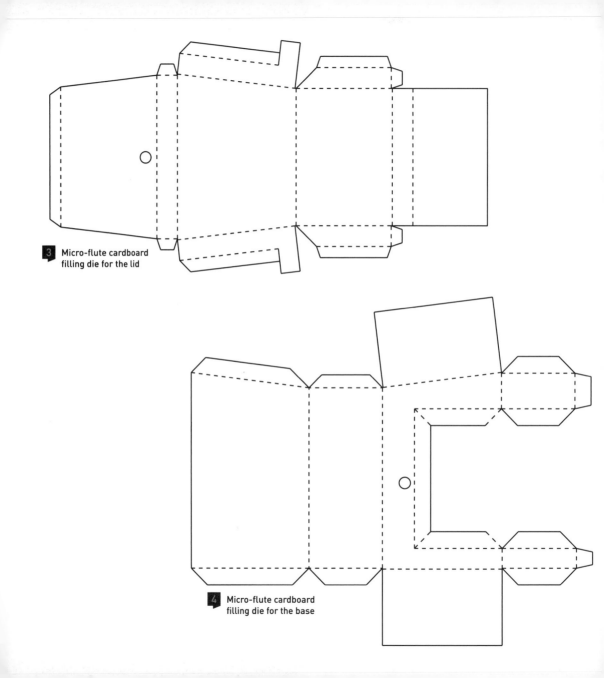

3 Micro-flute cardboard
filling die for the lid

4 Micro-flute cardboard
filling die for the base

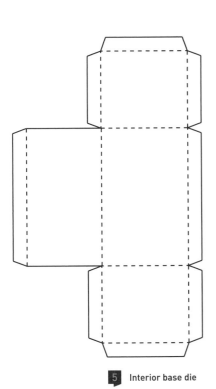

5   Interior base die

109

CREDIT CARD PACK

Use: Welcome pack. **Development:** A U-shaped box was designed with a thick lid and an additional flap where an information booklet goes. This lid has the same thickness as the sides of the U so that it adjusts perfectly to the base and allows for snapping the box shut. A 3 mm foam sheet is placed between the interior piece and the exterior piece to fix the card to the base. **Materials:** 300 g paperboard, 2 mm EVA foam and double-layer corrugated cardboard. → 1255

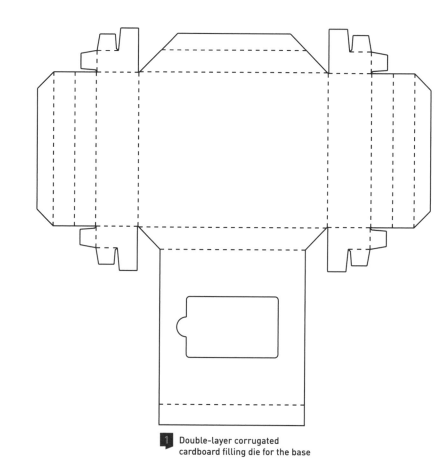

1 Double-layer corrugated cardboard filling die for the base

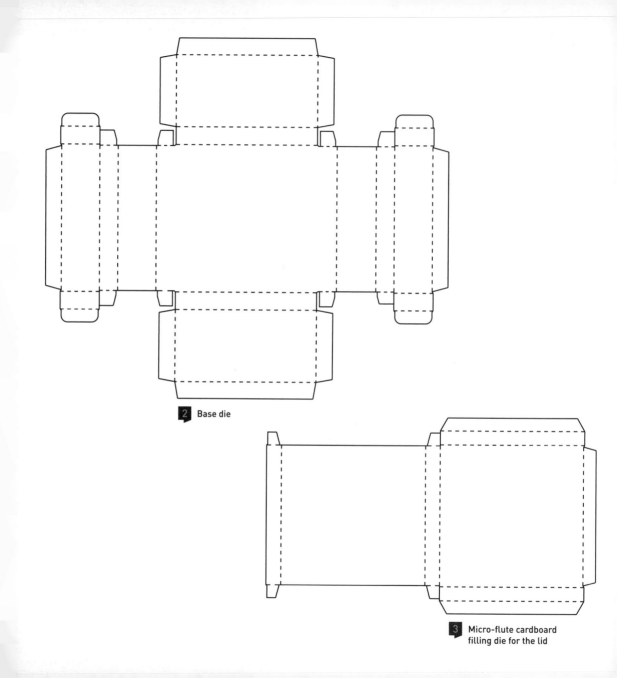

2 Base die

3 Micro-flute cardboard filling die for the lid

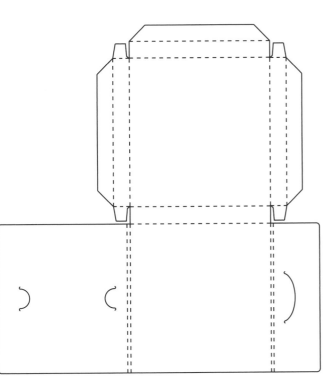

 Lid die

110

WOODEN CONTAINER BOX

Use: Box for presenting a product line via the press and influences.. Development:. Box made from cardboard and lined with balsa wood. It consists of a base with a mirror and a cover with elastic straps and metal clasps. Materials: 300g paparboard, 2mm gray carboard, 5 mm foam, 30mm rubberband, rivet fastener, 130g lining paper and bouble-layer corrugated cardboard and valsa wood, polyest er mirror silver. →2404

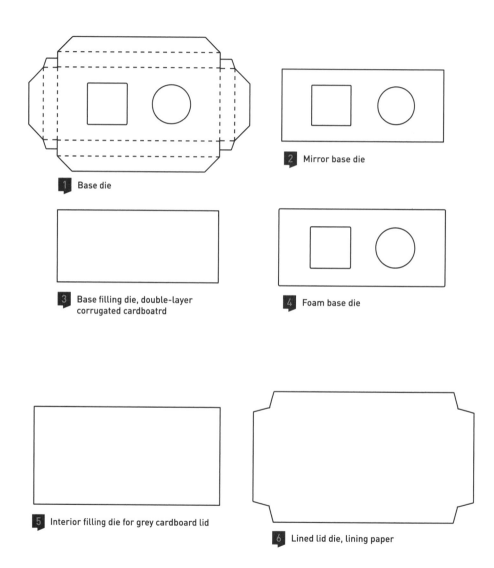

1 Base die

2 Mirror base die

3 Base filling die, double-layer corrugated cardboatrd

4 Foam base die

5 Interior filling die for grey cardboard lid

6 Lined lid die, lining paper

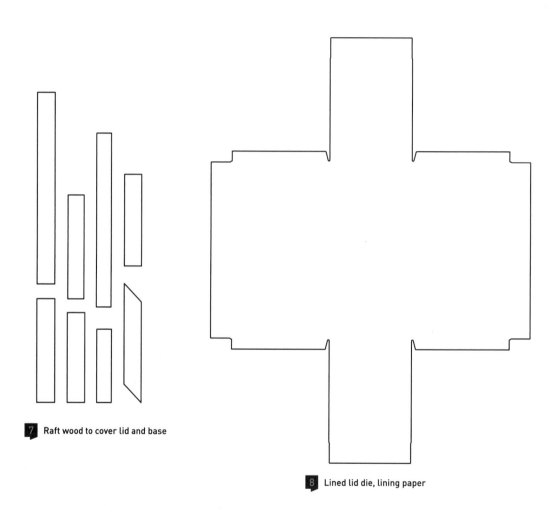

7 Raft wood to cover lid and base

8 Lined lid die, lining paper

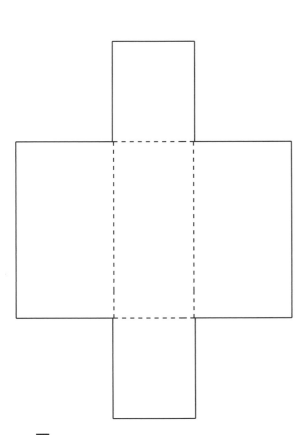

9 Intererior filling die for grey cardboard lid

111

PERSONALIZED FACIAL TREATMENT

Use: Premium display. **Development:** This packaging consists of 4 pieces: a base, two lids and a rigid exterior lid. The base is assembled with a interior piece of double-layer corrugated cardboard, to which cardboard hinges are added to support the lids. The exterior lid joined to the base serves as protection during shipment. One of the lids has a pocket that contains information. Interior magnets are used to ensure proper closing. **Materials:** 300 g paperboard, micro-flute cardboard, 3 mm elastic band, 18 mm wide satin ribbon, 2 mm gray cardboard. → 2372

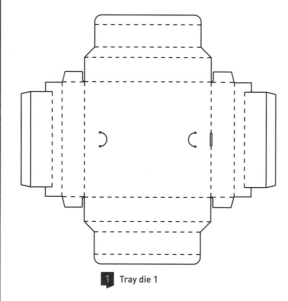

1 Tray die 1

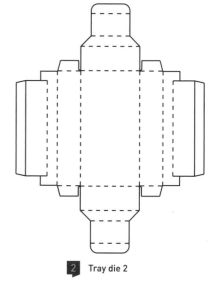

2 Tray die 2

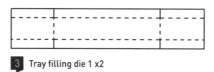

3 Tray filling die 1 x2

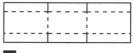

4 Tray filling die 2 x2

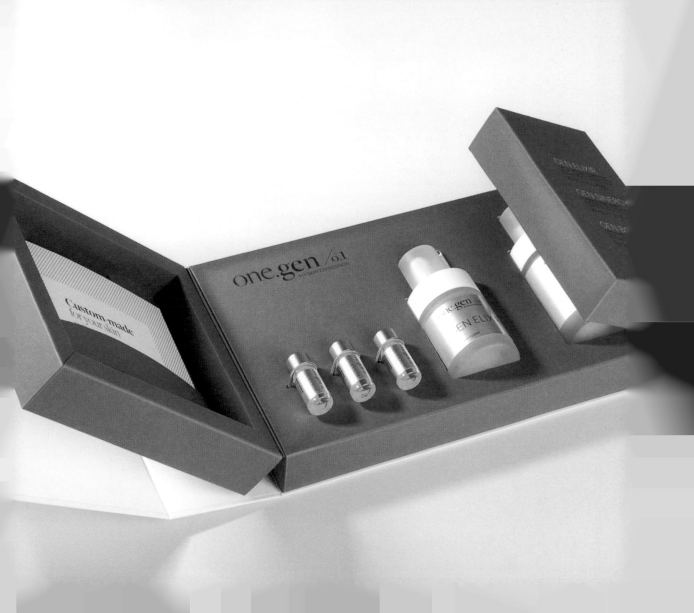

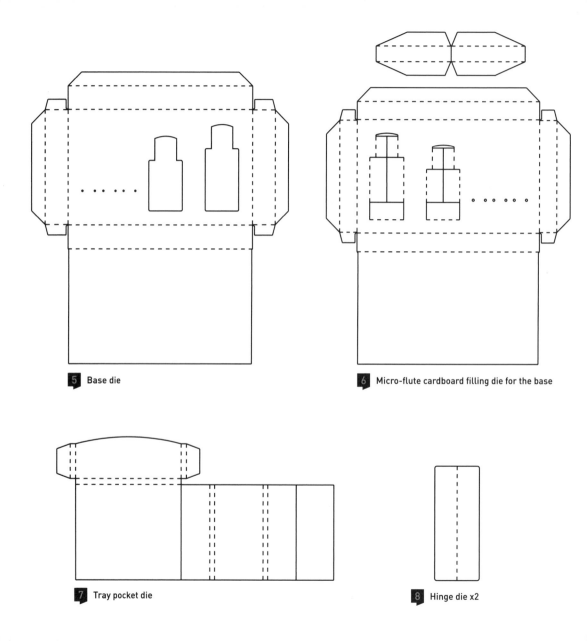

5 Base die

6 Micro-flute cardboard filling die for the base

7 Tray pocket die

8 Hinge die x2

 Lid die

112

RIGID TRIANGLE IN MOTION

Use: Display of content using new formats: **Development:** Triangle made with two pieces: a tray and a lid with interior pieces adapted to the form of the content. The lid is constructed with an interior piece of 2 mm gray cardboard. A cardboard hinge joins the pieces. For this reason, the height of the base must be kept in mind, as it cannot be overly tall due to the movement of the lid.
Materials: 300 g paperboard, micro-flute cardboard, 9 mm ø magnet and 12 mm wide satin ribbon. → 1898

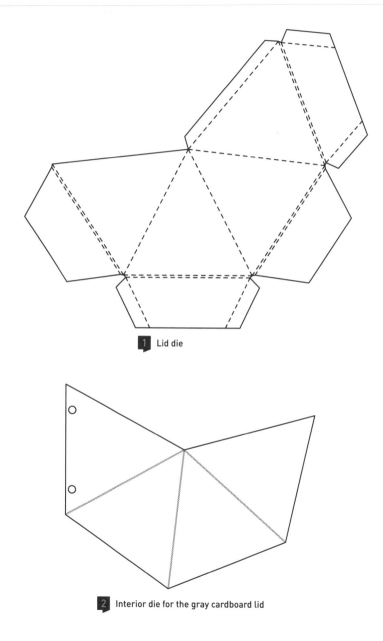

1 Lid die

2 Interior die for the gray cardboard lid

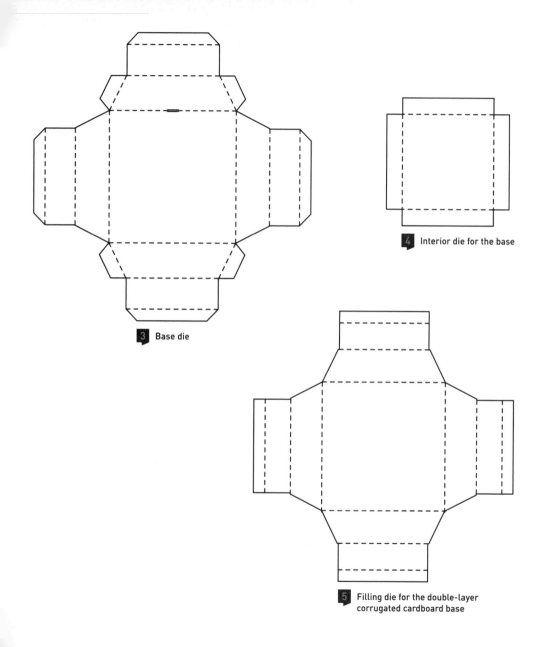

3 Base die

4 Interior die for the base

5 Filling die for the double-layer corrugated cardboard base

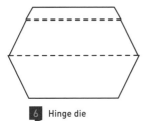

6 Hinge die

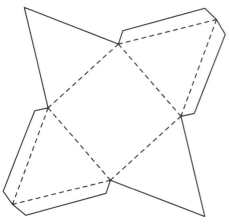

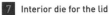

7 Interior die for the lid

113

DOUBLE FRAGRANCE WELCOME PACK

Use: Display of two fragrances in a single pack. **Development:** Tray to which two interior lids, a rigid exterior lid and an elastic-band closing mechanism are added. The interior of the tray is made with die-cut foam in the shape of the bottles to protect them, and a die was added to preserve the thickness of the elastic band. **Materials:** 300 g paperboard, gray cardboard, cloth paper, 30 mm foam, 20 mm elastic band, 20 mm ribbon. → 2325

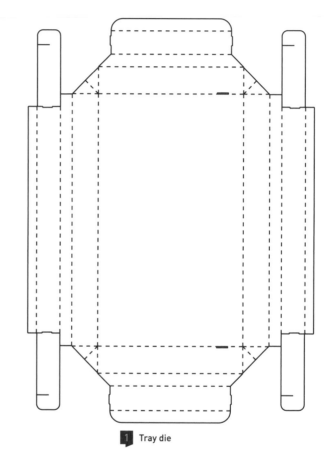

1 Tray die

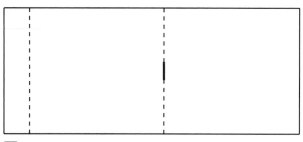

2 Interior die for the lids x2

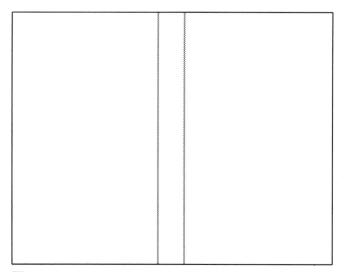

3 Filling die for gray cardboard lid

4 Filling die for gray cardboard lid

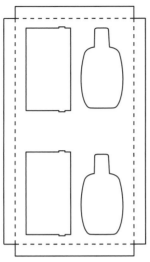

5 Interior die

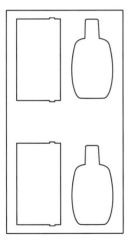
6 Filling die for interior foam

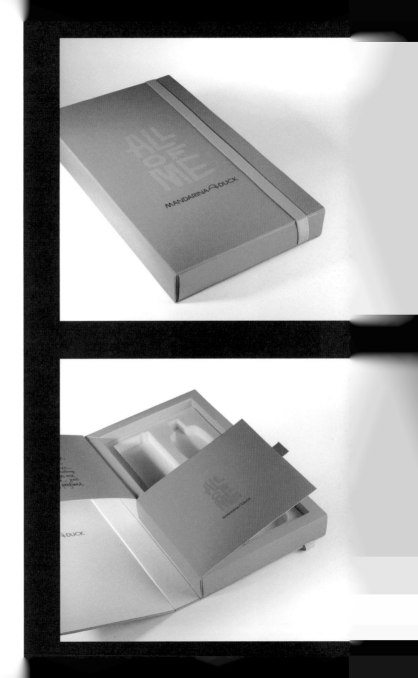

114

TRANSFORMABLE CUBE BOX

Use: Limited edition product display box. **Development:** This pack appears to be a conventional rectangular box though in reality it is formed by two symmetrical triangular pieces. If we fold these two pieces, the box becomes a pyramid-shaped display case exhibiting the products inside. It should be kept in mind that the products need to protrude enough so that they come into contact with the lid, thereby preventing them from moving when the pack is closed. **Material:** 300 g paperboard.→ 2323

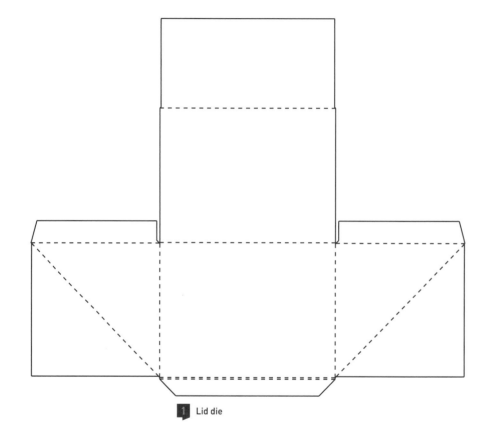

1 Lid die

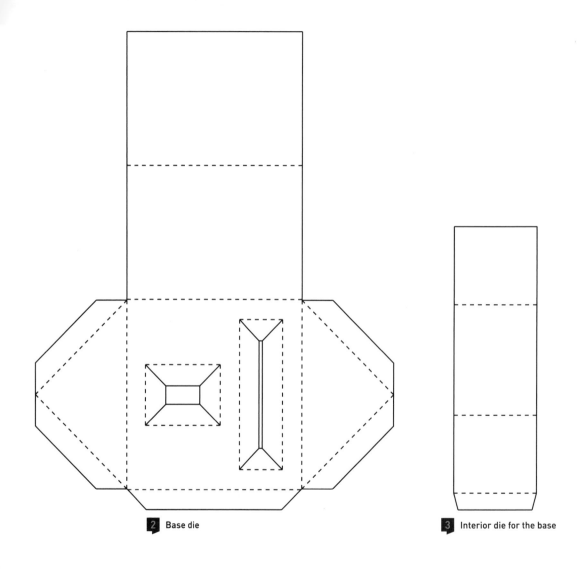

2   Base die

3   Interior die for the base

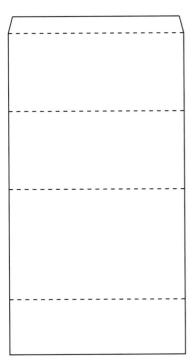

4 Interior die for the lid

115

BEWITCHING
DISPLAY BOX

Use: Product display box with a very feminine look. **Development:** This welcome pack consists of a black methacrylate base that blocks the lid and contributes an air of elegance. The base incorporates a die-cut foam piece in the shape of the products, which snap into place. This prevents the small products that don't touch the lid from moving when the pack is closed. **Materials:** 300 g paperboard, 3 mm methacrylate, 40 mm foam, 3 mm ø elastic band. → 2375

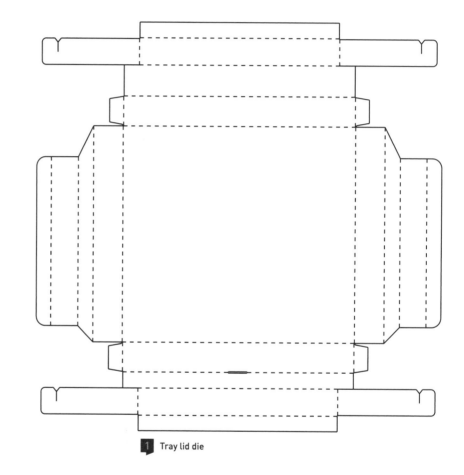

1 Tray lid die

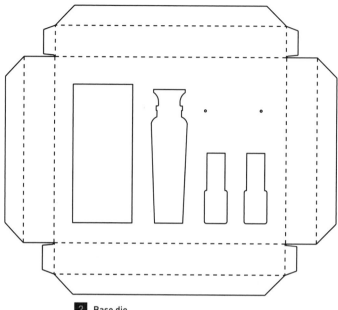

2 Base die

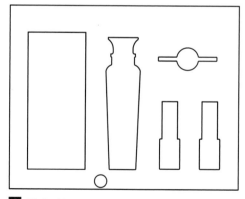

3 Die for 40 mm foam base

 Methacrylate plane and elevation

116

POPCORN
SURPRISE BOX

Use: Unique box with a product sample hidden inside. **Development:** This rectangular box conceals a tasty interior. It is formed by two pieces that fit together. The outer piece is a perforated frame that holds a transparent box containing a product sample. This exterior piece also has a flap that serves as an informative diptych. **Materials:** 300 g paperboard, 0.3 mm PET plastic. → 1437

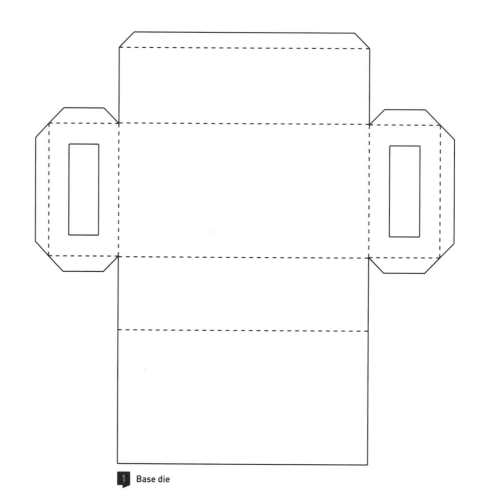

1 Base die

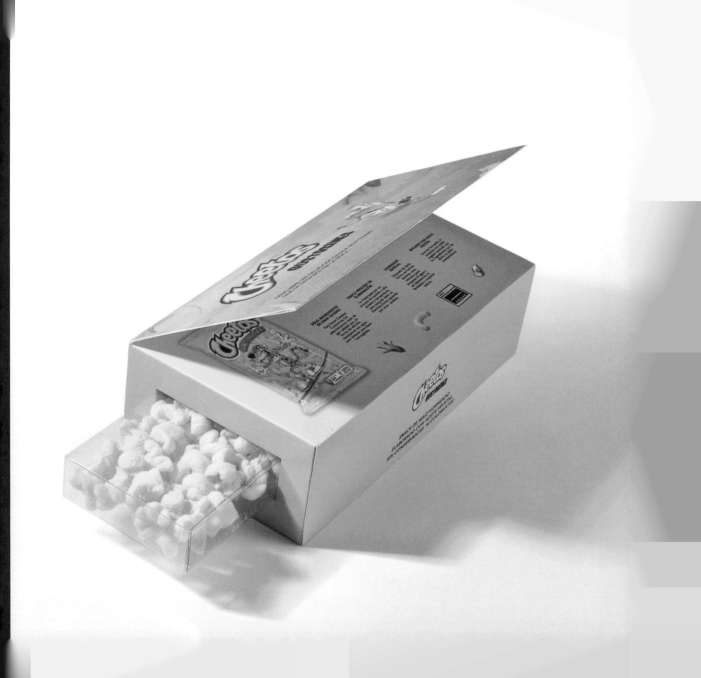

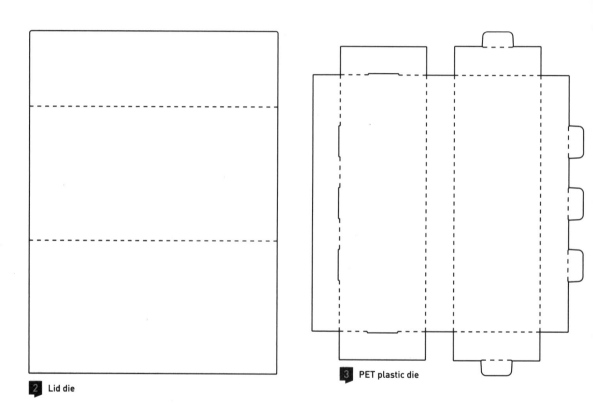

2 Lid die

3 PET plastic die

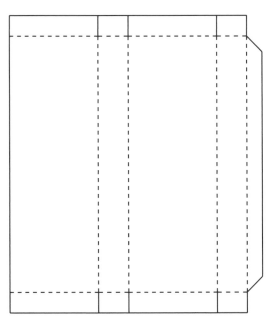

4 Interior base die

117

PASSE-PARTOUT BOOK PACK

Use: Book display. **Development:** Made of unconventional materials, this box incorporates wood, leather, felt and methacrylate to cover the product. The box consists of a wooden base and several pieces of felt (passe-partout) that protect it. The methacrylate lid has a slot in the frame so that its placement on the felt is centered and a leather strap that attaches the book to the box and functions as a closing mechanism. **Materials:** 10 mm felt, wood, methacrylate, leather belt. → 0568

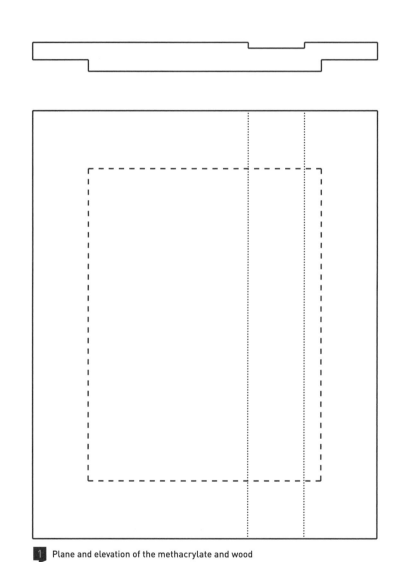

1 Plane and elevation of the methacrylate and wood

DE L'ESPRIT
DES
LOIX
"MONTESQUIEU"
ANY 1749
1ERA EDICIÓ

 Felt die x8

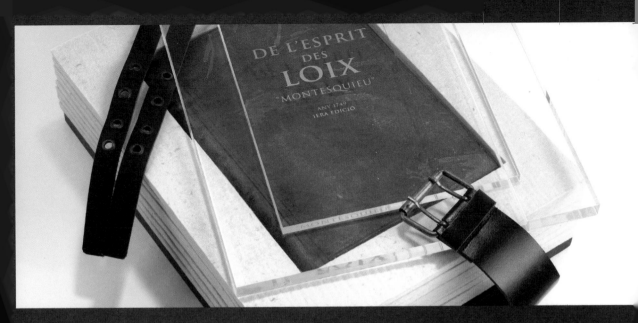

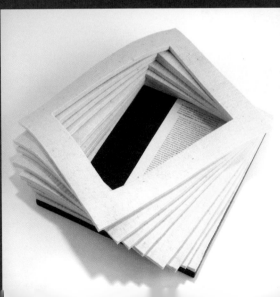

118

NEW
PERFUME PACK

Use: Welcome pack. **Development:** This packaging consists of 2 pieces: the base as the main element and a lid embedded inside of it. The lid is created from a conventional tray joined at the base by a cardboard hinge. The interior of the base has been reinforced with die-cut foam filling to protect the contents. Special care needs to be taken regarding the thicknesses of the lid and the hollow interior space to ensure proper closing. **Materials:** 300 g paperboard, double-layer corrugated cardboard, 40 mm foam, 3 mm ⌀ elastic band. → 2352

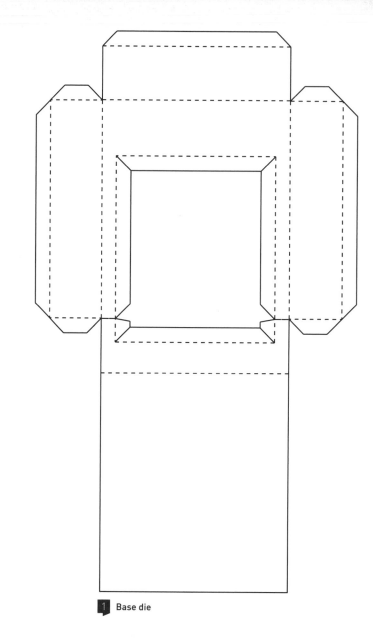

1 Base die

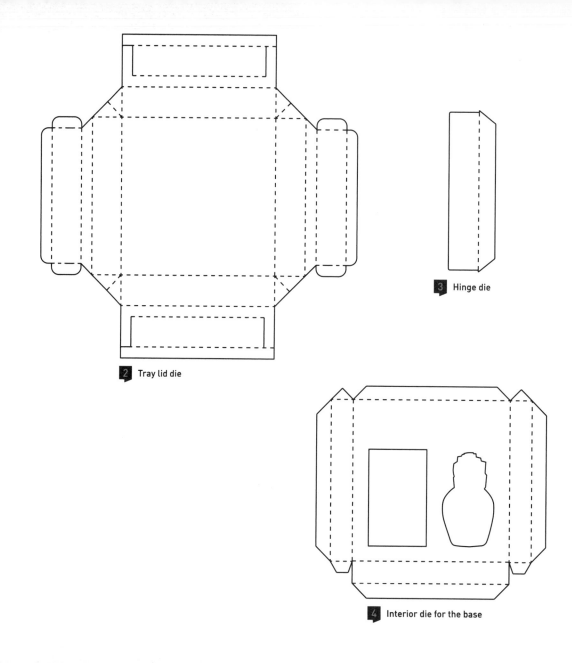

2 Tray lid die

3 Hinge die

4 Interior die for the base

5 Filling die for the double-layer corrugated cardboard base

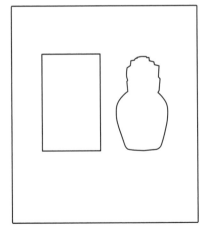

6 Filling die for the 40 mm foam base

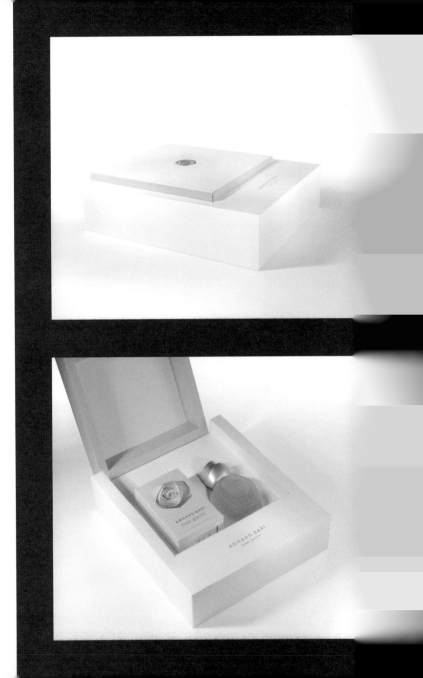

BOOK
FRAME

Use: Welcome pack. **Development:** This pack consists of a thick sleeve of reinforced paperboard with micro-flute cardboard inside of which we have placed a book. The thickness of the sleeve allows for creating a compartment to hold the product, as we have done here, covering it with an outer flap that closes with the help of interior magnets. **Materials:** 300 g paperboard, micro-flute cardboard, 10 mm ø magnet and strapping band. → 1983

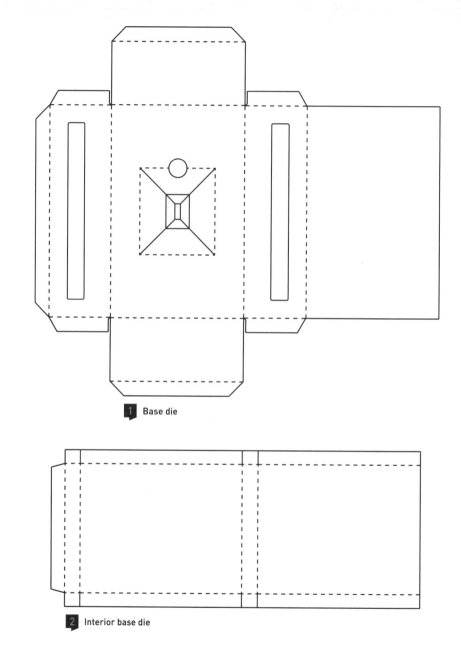

1 Base die

2 Interior base die

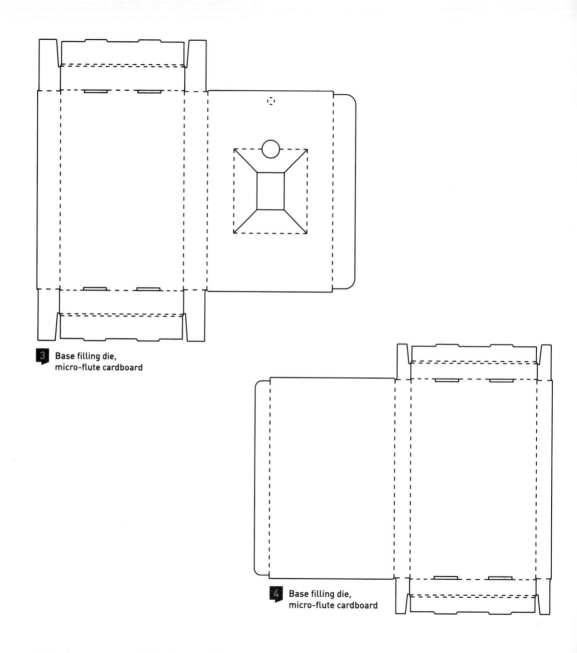

3 Base filling die,
micro-flute cardboard

4 Base filling die,
micro-flute cardboard

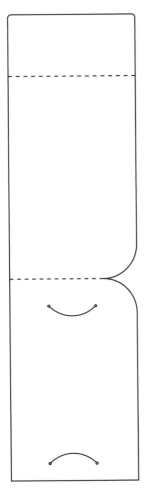

5 Lid die

SLOPED
WELCOME PACK

Use: New fragrance display. **Development:** The aesthetic of this box evokes the material out of which the product packaging that contains the wood is made. The cardboard pack is structured in three levels, creating the sensation of a series of rectangular wooden pieces stacked in a "carefully chaotic" manner. The finishing of the pack has a rustic and natural look, highlighting the grain of the wood. **Materials:** 300 g paperboard, 20 mm foam, 3 mm ø elastic band. → 2175

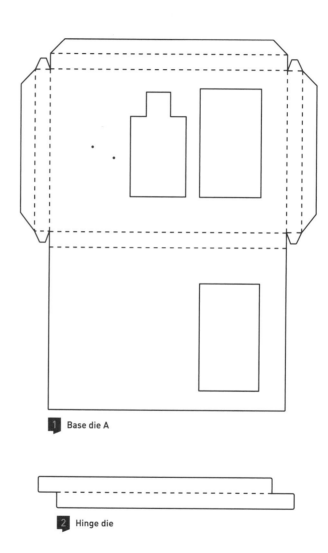

1 Base die A

2 Hinge die

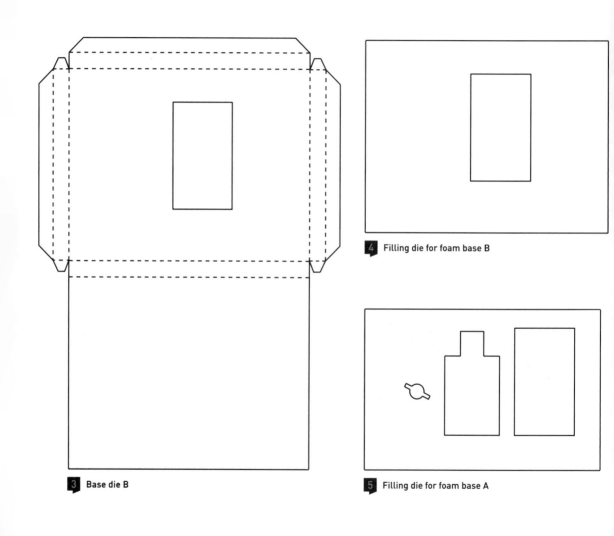

3 Base die B

4 Filling die for foam base B

5 Filling die for foam base A

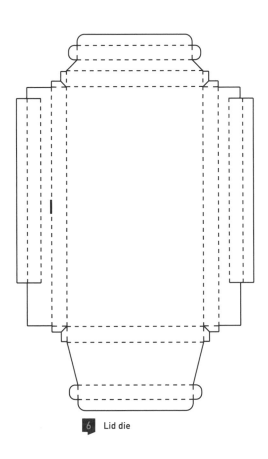

6 Lid die

121

PADDED DISPLAY CASE FOR GLASSES

Use: Glasses display case. **Development:** This box is made with an L-shaped methacrylate base and lid. When the lid is opened, it remains in a vertical position for display. The red methacrylate includes a transparency that reveals a printed sheet positioned in the form of a second lid. The inside is made out of foam, with a system of nests that protect the product. **Materials:** 300 g paperboard, 5 mm methacrylate. 25 mm foam. → 1890

1 Methacrylate plane and elevation

2 Foam base die A

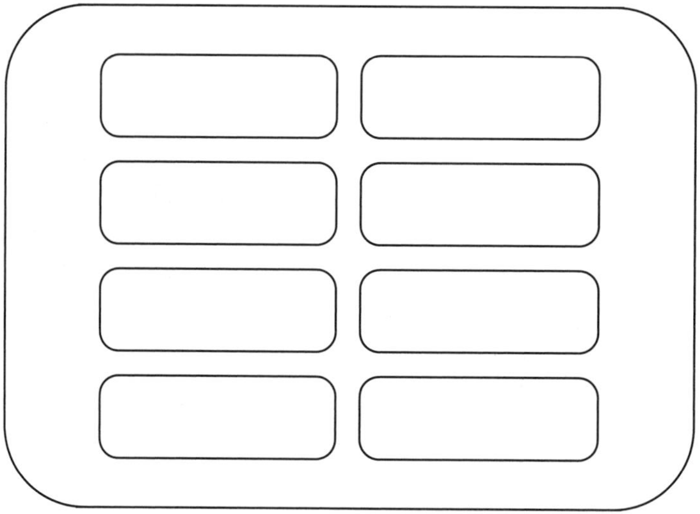

3 Foam base die B

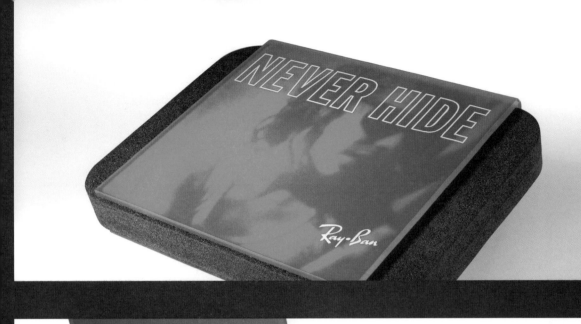

122

TRIPLE
LID BOX

Use: Individual product display. **Development:** This display box contains three objects arranged on the same perforated base. The interior thickness of the lids must be the same as the part that juts out from the product to ensure the latter remains secure when closing the pack. The exterior thickness of the trays must be the same as the base so that when the pack is open, the base and lid are balanced. **Materials:** 300 g paperboard, 10 mm magnet and grey cardboard. → 1226

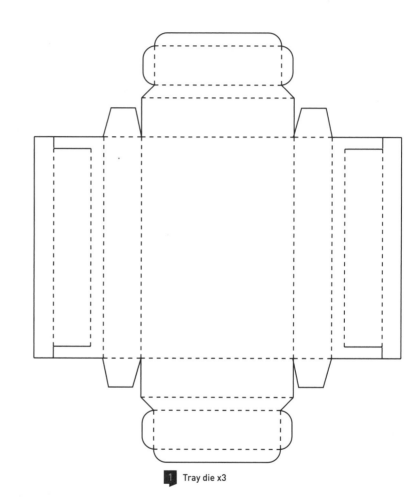

1 Tray die x3

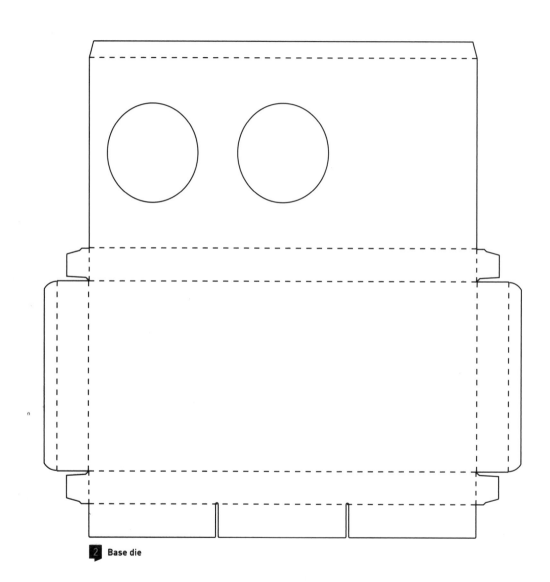

2 Base die

3 Interior die for grey cardboard tray x3

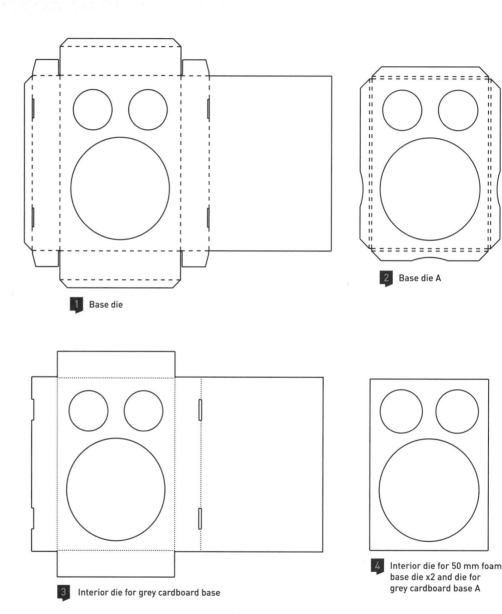

123

SOMMELIER GIFT PACK

Use: Gift case for bottles and glass jar. **Development:** A lid and a perforated base form from this pack for containing the fragile objects affixed to the base and held in place by an interior foam piece. The lid is a tray that covers the product and base of the box completely, and includes two elastic straps to ensure proper closing of the pack. **Materials:** 300 g paperboard, 2 mm gray cardboard, 50 mm foam, 30 mm rubber band, rivet fastener, 130 g lining paper and double-layer corrugated cardboard. → 1669

1 Base die

2 Base die A

3 Interior die for grey cardboard base

4 Interior die for 50 mm foam base die x2 and die for grey cardboard base A

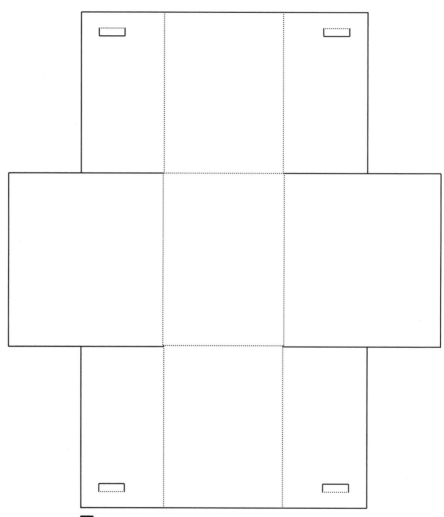

5 Interior filling die for grey cardboard lid

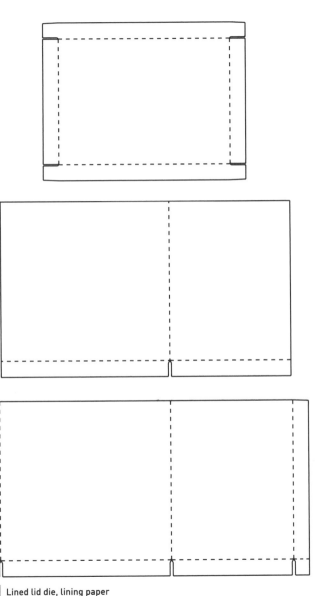

6 Lined lid die, lining paper

WOODEN DISPLAY FRAME

Use: Display pack. **Development:** This elegant box was designed to cover a wooden display frame. It is formed by two L-shaped pieces of cardboard that envelop the frame, revealing its sides. The lid is made out of laid cardboard with the same gold design as the upper part of the bottle. The box has an interior piece in the exact shape of the bottle to protect it: **Materials:** 300 g paperboard, 50 mm foam, 25 mm wood. → 1228

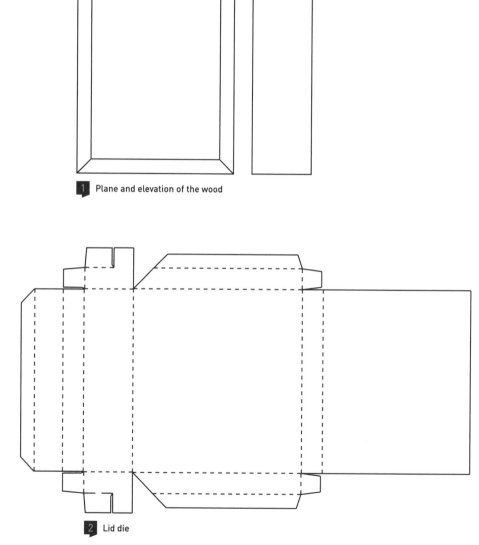

1 Plane and elevation of the wood

2 Lid die

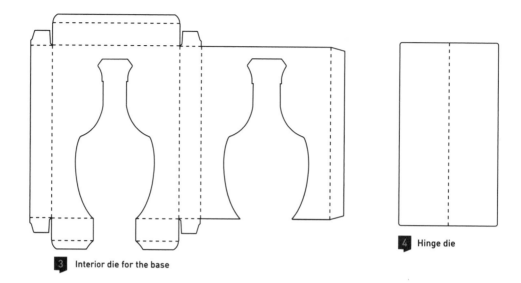

3 Interior die for the base

4 Hinge die

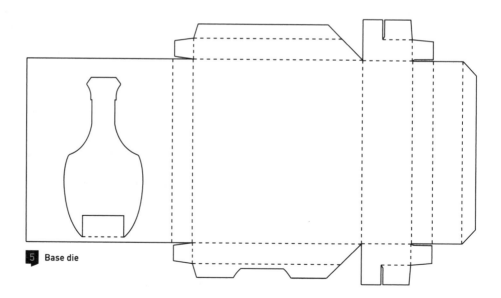

5 Base die

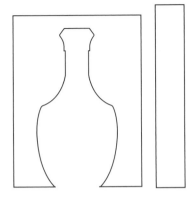

6 40 mm foam die

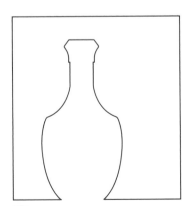

7 30 mm foam die

LUMINOUS
WELCOME PACK

Use: Box with LEDs and a revolving piece to highlight the product. **Development:** Original pack formed by an outer frame and a revolving inner frame that calls attention to the product inside. in this case, a bottle. When the pack is opened, an LED lighting system illuminates the bottle automatically. **Materials:** 300 g paperboard, micro-flute cardboard, 5 mm methacrylate, double-layer corrugated cardboard, 18 mm magnet, 10 mm ø EVA foam, LED lighting system (battery, LEDs, switch and cables). → 1746

1 Sleeve die

2 Lid die

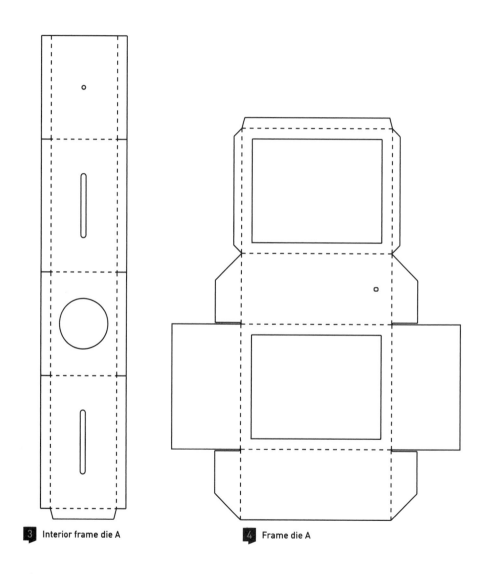

3 Interior frame die A

4 Frame die A

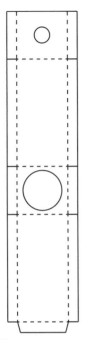

5 Interior frame die B

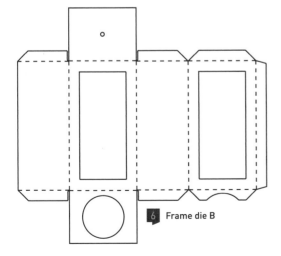

6 Frame die B

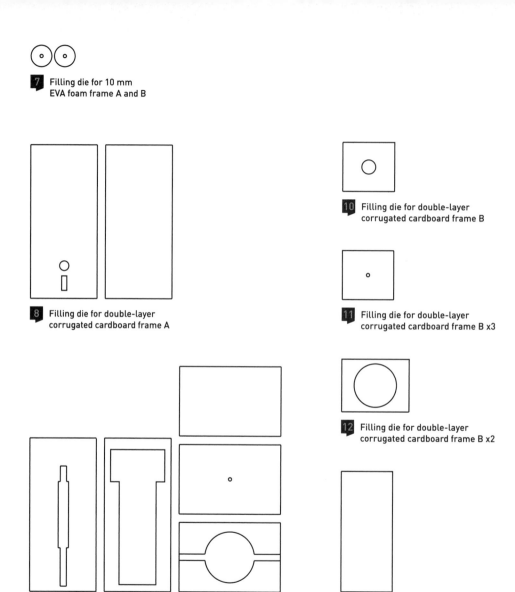

7 Filling die for 10 mm
EVA foam frame A and B

8 Filling die for double-layer
corrugated cardboard frame A

9 Filling die for double-layer
corrugated cardboard frame A

10 Filling die for double-layer
corrugated cardboard frame B

11 Filling die for double-layer
corrugated cardboard frame B x3

12 Filling die for double-layer
corrugated cardboard frame B x2

13 Filling die for double-layer
corrugated cardboard frame B x6

JOSEP MARIA
GARROFÉ

JOSEP Mª GARROFÉ (BARCELONA, 1967) WAS ORIGINALLY DRAWN TO THE PACKAGING WORLD BY THE CHALLENGE IT PRESENTED OF FORGING NEW CREATIVE PATHS THAT DEPARTED FROM CONVENTIONAL THINKING. AT THE AGE OF 23, HE FOUNDED HIS OWN STUDIO, **GARROFÉ BRAND&PACK**, WHICH QUICKLY GAINED TRACTION PROVIDING RIGOROUS STANDARDS AND A HIGH LEVEL OF FORMAL INNOVATION TO ALL OF HIS PROJECTS. NINE YEARS LATER, **TRIBU-3** WAS CREATED WITH THE AIM OF OFFERING COMPREHENSIVE CUSTOMER SERVICE IN THE PRODUCTION OF ALL THE CLIENT'S DESIGNS. FINALLY, IN 2010, **SELFPACKAGING.ES**, AN INNOVATIVE ONLINE PACKAGING PROJECT WITH A STRONG INTERNATIONAL FOCUS THAT OFFERS USERS THE POSSIBILITY OF CHOOSING AND CUSTOMIZING THEIR OWN PACKS, EMERGED ON THE SCENE.

THE INNATE NEED TO OVERCOME BARRIERS AND CONTINUE SEARCHING FOR NEW CHALLENGES ALSO COMPELLED JOSEP TO PARTICIPATE IN THE DAKAR RALLY ON SEVERAL OCCASIONS IN THE SEARCH FOR ANSWERS THAT CAN ONLY BE FOUND IN THE IMMENSITY OF THE DESERT, HIS GREAT PASSION. THE DESIRE TO PRODUCE THE HIGHEST QUALITY WORK AND DISSEMINATE IT FOR THE ENJOYMENT OF OTHERS IS THE SEED OUT OF WHICH HIS PREVIOUS PUBLICATIONS **STRUCTURAL DISPLAYS, STRUCTURAL PACKAGING, REFUSE AND STRUCTURAL GREETINGS,** WHICH QUICKLY BECAME BESTSELLERS, GREW. AND IT IS FOR THIS SAME PURPOSE THAT JOSEP NOW PRESENTS **NEW STRUCTURAL PACKAGING GOLD**, A SELECTION OF PROJECTS THAT ILLUSTRATE AND SYNTHESIZE HIS LEARNING, EXPERIENCE AND UNFLINCHING ADVENTUROUS SPIRIT.

Specially for laura

THANK YOU

/SPG/

THIS BOOK WOULD NEVER HAVE BEEN POSSIBLE WITHOUT THE DETERMINATION AND DEDICATION OF LAURA FARRÉ. SHE IS THE BOOK'S SOUL, THE ONE WHO PAMPERED IT AND ALLOWED IT TO GROW INTO THE PUBLISHING SUCCESS THAT UNDOUBTEDLY IT WILL SOON BECOME. I TAKE GREAT PRIDE IN HER INVALUABLE COLLABORATION, WHICH GIVES ME THE PEACE OF MIND OF KNOWING THAT EVERYTHING WILL TURN OUT WELL. I WOULD LIKE TO TAKE THIS OPPORTUNITY TO OFFER HER MY SINCEREST THANKS AND TO TELL HER THAT I HOPE SHE STILL HAS ENOUGH PATIENCE LEFT TO ACCOMPANY ME ON FUTURE PUBLISHING ADVENTURES.

THE EFFORTS OF THE DESIGN AND PRODUCTION TEAM AT **GARROFE.COM** AND **TRIBU-3.COM** WERE EQUALLY INSTRUMENTAL IN ENSURING THAT THE BOOK REACHED A SAFE HARBOR.

LAST BUT NOT LEAST, I WISH TO EXPRESS MY DEEP GRATITUDE TO ALL OUR CLIENTS FOR THEIR PATIENCE AND COURAGE IN ACCEPTING THE CHALLENGES THAT WE HAVE SET BEFORE THEM IN THE FORM OF OUR IDEAS. WITHOUT THE TRUST THAT THEY PLACE IN US EACH AND EVERY DAY, **NEW STRUCTURAL PACKAGING GOLD** WOULD HAVE HAD A DIFFICULT TIME EVER SEEING THE LIGHT OF DAY.

LEVEL
/SPG01/

LEVEL
/SPG02/

LEVEL
/SPG03/

002_Star-shaped surprise box 16
032_Christmas tree pack 76
035_Booster pack 82
042_Gift box with cord 96
048_Gift bag with handle 108
059_Two interweaved triangles 144
075_Product on display 198
087_Perforated square pack 244
098_Christmas tree box 288

018_Document holder 48
051_Software triptych 120
052_Double-layer cardboard sleeve 124
061_Block for CD 150
062_Two tray pack 152
063_Book with grass 156
068_Grass between blocks 174
086_Gift kit sample case 240
112_Rigid triangle in motion 344
117_Passe-partout book pack 364

056_Curved plastic packaging 136
102_Display tray insert 304
107_Pack with curved corners 324
109_Credit card pack 332

014_Eva foam bag 40
025_Arched base box 62
046_Surprise pack 104
074_Ingot-shaped box 196
077_Connected pack package 204
084_Double symmetry box 232
121_Padded display case for glasses 380

010_Rabbit face pillow box 32
016_Elephant candy box 44
021_Halloween bat with eyes 54
022_Candy cone 56
023_Four triangle box 58
029_Box for food 70
044_Matryoshka box 100

003_Picnic box/bag 18
015_Chinese noodle box 42
017_Double sleeve cube 46
039_Surprise box with "Ears" 90
064_Multi-use box-bag 160

004_Wedding gift box 20
006_Heart box 24
012_Simple rounded bag 36
013_Triangular case 38
045_Two embedded triangles 102

079_Triptych display box 212
089_Car welcome pack 252
092_"L"-border box 264
103_Plush welcome pack 308
106_Globe cube 320
119_Book frame 372
120_Sloped welcome pack 376
124_Wooden display frame 392
125_Luminous welcome pack 396

001_Rigid accordion bag 14
028_Carrot box 68
033_Cardboard heart for bottle 78
053_Cylinder for credit cards 126
065_Multi-layer CD triptych 164
067_Note pad folder 172
073_Advent calendar box 192
080_Ecological Christmas pack 216

019_Rigid tray with handles	50
026_Simple cupcake box	64
027_Cup holder structure	66
031_Desserts take away box	74
071_Box with raised base	184
099_Coffee maker display pack	292
116_Popcorn surprise box	360
122_Triple lid box	384

005_Pillow box	22
007_Triangle surprise	26
011_Double pyramid box	34
020_Valentine's Pack	52
024_Box for cakes	60
047_Tied-together triangles	106
050_Welcome pack with chocolates	116
057_Six triangle cube	138
069_Three layer pack	178

038_Perfume diptych	88
058_Multi-layer sample box	140
081_Two interlocking "L"s	220
083_Minimalist display case	228
097_Display tray welcome pack	284
100_Methacrylate box	296
110_Wooden container box	336
111_Personalized facial treatment	340
114_Transformable cube box	352
118_New perfume pack	368

009_Portable belt for flowers	30
030_Flower basket	72
034_Ball inserted in a cross	80
037_Batman box	86
043_Car-shaped box	98
088_Framed gift box	246
094_Diagonal surprise pack	272
101_Promotional case pack	300
108_Sloped box	328

008_Double front package	28
040_Foam box for usb	92
082_Foam box with handle	224
055_Interlinking sleeves pack	132
085_Double cube pack	236

036_Four-bottle structure	84
041_Triangle with triangular closure	94
049_"Twisted" box for bottles	112
054_Foam triptych box	128
060_Two-color surprise box	146
070_Gift box for bottles	182
076_Sommelier display pack	202
090_Rounded-edge chest	256
105_Encased bottle pack	316
123_Sommelier gift pack	388

066_Triple display box	168
072_Gift case with sleeve	188
078_Triple bucket pack	208
091_Mobile display frame	260
093_Multiple accordion box	268
095_Fragrance welcome pack	276
096_New fragrance display	280
104_Women's perfume chest	312
113_Double fragrance welcome pack	348
115_Bewitching display box	356

INDEX THEM

www.structuralpackagingblog.com

STRUCTURAL PACKAGING BLOG IS A SITE FOR THE EXCHANGE OF IDEAS, WHERE YOU'LL FIND FRESH AND INNOVATIVE CONTENT ABOUT ISSUES RELATED TO THE AREA OF STRUCTURAL PACKAGING

The idea of developing the blog is the result of the increasing need to contribute and share our view and knowledge within the sector with all those people who are interested in the world of structural packaging.

The Garrofé Agency website, **www.garrofe.com**, and SelfPackaging **www.selfpackaging.com** complement the blog.

www.garrofe.com

GARROFÉ IS A DESIGN AGENCY AND PACKAGING ATELIER

Founded in Barcelona in 1991 by Josep M. Garrofé. We share our passion and discipline, designing packaging that seduces and connects with people and brands. Our goal is designing pieces that are unique in materials, shape, structure, finishes and quality.

www.selfpackaging.com

SELFPACKAGING IS OUR ONLINE SHOP

It was born with the goal of developing packaging that is accessible to everybody, with no minimum order quantity, affordable and with a catalogue full of multiple shapes, accessories and tools to support creativity and professional production.

STRUCTURAL PACKAGING
/BLOG/